ROUTLEDGE LIBRARY EDITIONS:
POLLUTION, CLIMATE AND CHANGE

Volume 12

TOWARDS EFFICIENT REGULATION OF AIR POLLUTION FROM COAL-FIRED POWER PLANTS

TOWARDS EFFICIENT REGULATION OF AIR POLLUTION FROM COAL-FIRED POWER PLANTS

ROBERT O. MENDELSOHN

Routledge
Taylor & Francis Group

LONDON AND NEW YORK

First published in 1979 by Garland Publishing, Inc.

This edition first published in 2020
by Routledge
2 Park Square, Milton Park, Abingdon, Oxon OX14 4RN

and by Routledge
52 Vanderbilt Avenue, New York, NY 10017

Routledge is an imprint of the Taylor & Francis Group, an informa business

© 1979 Robert O. Mendelsohn

British Library Cataloguing in Publication Data
A catalogue record for this book is available from the British Library

ISBN: 978-0-367-34494-8 (Set)
ISBN: 978-0-429-34741-2 (Set) (ebk)
ISBN: 978-0-367-36776-3 (Volume 12) (hbk)
ISBN: 978-0-367-36777-0 (Volume 12) (pbk)
ISBN: 978-0-429-35138-9 (Volume 12) (ebk)

Publisher's Note
The publisher has gone to great lengths to ensure the quality of this reprint but points out that some imperfections in the original copies may be apparent.

Disclaimer
The publisher has made every effort to trace copyright holders and would welcome correspondence from those they have been unable to trace.

Towards Efficient Regulation of Air Pollution From Coal-Fired Power Plants

Robert O. Mendelsohn

Garland Publishing, Inc.
New York & London, 1979

Library of Congress Cataloging in Publication Data

Mendelsohn, Robert O   1952-
  Towards efficient regulation of air pollution from coal-fired
power plants.

  (Outstanding dissertations on energy) (Outstanding
dissertations in economics)
  Originally presented as the author's thesis, Yale.
  Bibliography: p.
  1. Steam power-plants—Environmental
aspects—Mathematical models.   2. Air—Pollution—
Mathematical models.   I. Title.   II. Series.   III. Series:
Outstanding dissertations in economics.
TD888.S7M46   614.7'12   78-75020
ISBN 0-8240-4055-4

All volumes in this series are printed on acid-free,
250-year-life paper.

Printed in the United States of America.

TOWARDS EFFICIENT REGULATION OF AIR POLLUTION
FROM COAL-FIRED POWER PLANTS

A Dissertation

Presented to the Faculty of the Graduate School

of

Yale University

in Candidacy for the Degree of

Doctor of Philosophy

by

Robert Owen Mendelsohn

December 1978

Acknowledgements

I am indebted to Professor Guy Orcutt who has generously offered his time, energy, and professional interest over the last few years. His creative pursuit of knowledge, his dogged skepticism of unsupported but accepted ideas, and his high standards for economic research have earned him both my respect and admiration. I would also like to thank Susan Rose Ackerman whose criticisms have been both thorough and judicious. Professor Robert Dorfman of the Harvard Economics Department has been most helpful with his critical assistance at important stages of this research. I would also like to thank Professor William Nordhaus whose encouragement and support helped launch this research effort.

This multidisciplinary project could not have been completed without the assistance of several individuals outside of economics. I would like to thank James Crowe of United Illuminating and Paul Nordine of the Yale Engineering Department for their guidance with respect to air pollution control engineering. For their helpful explanations of meteorology and atmospheric chemistry, I would like to thank Ronald Meyers of Brookhaven National Laboratory, Mike MacCracken and Ken Peterson of Lawrence Livermore Laboratory, William Reifsnyder of the Yale School of Forestry and Environmental Studies, and Mike Ruby of the University of Washington. I would also like to thank Ronald Wyzga of the Electric Power Research Institute for sharing his knowledge of pollution dose response curves.

I especially want to thank Susan Watson for her able management of the final presentation of this manuscript.

This work was completed with the financial support of the National Science Foundation through a dissertation research support grant.

TABLE OF CONTENTS

LIST OF TABLES

LIST OF ILLUSTRATIONS

CHAPTER I

INTRODUCTION

Economic theory provides a simple but powerful insight into efficient pollution control: it saves money to minimize the sum of abatement costs and pollution damages. Applied economists, rarely able to estimate pollution damages, have not been in a position to realize even this basic insight. Environmental policy makers, eager to react to a pressing social problem, have been forced to design policies without a clear understanding of their impacts. Billions of dollars are currently being spent on pollution control in addition to losses due to the prohibition of many industrial, commercial, and residential activities. What is society buying with all these regulations? Can this environmental product be purchased more cheaply? Who benefits from different regulations? Should environmental regulations be altered on behalf of specific constituents?

Economists are just beginning to build models which incorporate both economic insights into externalities and appropriate environmental submodels. Though the model developed in this essay deals only with air pollution from coal fired power plants, it broadly illustrates how available scientific information can be organized to improve our understanding of pollution control. Armed with this information, applied economists can finally discuss the relevant consequences of specific air pollution abatement strategies. This improved understanding should encourage better decisions in a number of ways. First, the information should enable policy makers to discern and thereby avoid inferior abatement techniques. Second, the information is presented in such a way as to encourage conscious recognition of the value judgments which society must make in order to choose among the remaining efficient abatement strategies. This will make the implicit value judgments of every policy recognizeable to society and, hopefully, assure that these judgments are acceptable to society. Third, the distributional impacts of particular strategies can be better understood so that they can play an appropriate role in final policy decisions. Fourth, the model provides several empirical insights into preferable policy tools for controlling air pollution. Fifth, the reliability of different control techniques can be examined with respect to various aspects of air pollution control which are poorly understood.

A computer based environmental model is used in order to simulate the consequences of air pollution control measures upon society. These simulations illustrate the connection between abatement methods and their final environmental consequences. In order to demonstrate the usefulness of this approach, the model is applied to a specific case study. The object of this case study is the control of air pollution from a new, coal-fired, electrical generating station in New Haven. Further applications of this model to other sources and across more sites should be able to provide even more powerful policy insights. Far more than simply analyzing a part of the air pollution problem of one town, this research advances applied risk analysis by combining the insights of economics and environmental sciences.

The remainder of this chapter reviews the history and current status of the literature on both theoretical and applied air pollution modelling. The second chapter provides an introduction and outline to the environmental/economic model. The following three chapters contain technical descriptions of important segments of the model. The technical areas covered are abatement engineering, atmospheric modelling, and dose response functions. The sixth chapter introduces the simulations which are made using the environmental model. The seventh chapter discusses the effectiveness of various abatement techniques and the available tradeoffs for this case study. The eighth chapter examines the distributional impacts of different strategies and explores tradeoffs between efficiency and various forms of equity. The final chapter discusses the limitations of the study, the implications for public policy, and possible extensions or improvements of the analysis.

CHAPTER I-A

REVIEW OF THE ABSTRACT ISSUES[1]

Economic theorists have long been concerned about pollution partially because of its noticeable impact on human welfare, but primarily because emissions are not regulated by the "invisible hand" of the market place. As long ago as 1920, Pigou expressed his concern about smoke from chimneys because it "... inflicts a heavy uncharged loss on the community, in injury to buildings and vegetables, expenses for washing clothes and cleaning rooms, expenses for the provision of artificial light, and in many other ways."[2] It is not smoke, per se, that Pigou is concerned about, but rather the way that smoke is not controlled. Private interests, operating in a laissez-faire market system, are not charged for the damages they impose on the community by polluting common resources, such as the atmosphere or public water bodies. The polluter treats the resource as though it were free even though his use has very real costs to society. The effluents are "externalities" as long as the polluter is unconcerned about all the social costs he is imposing by producing them. Private and social interests diverge in this case, calling for market intervention by society or the government. The ideal intervention reconciles private and social net products so that the private party will act in the best interests of society. Pollution, to an economist, is under control when economic agents take the total environmental impact of their actions into account. Note that this will not necessarily lead to an abolition of pollution damages.

Though Pigou's insights into environmental economics have been invaluable, there have been significant refinements of his theories in

the last two decades. Coase [14], in 1960, started an important
discussion about the likelihood that the market would correct an
externality by itself. Using several court cases as example, Coase shows
that with costless negotiation, two parties can arrive at an efficient
solution to an externality without government intervention. Coase
points out that it would be in the victim's interest to offer the
polluter a bribe to curtail pollution as long as the bribe for a small
reduction is less than the marginal damage. Especially between two firms
(where there is no income effect), negotiations should result in a
situation where the marginal cost of abatement equals the marginal
damages. Coase concludes his argument by stating that with negotiation
the market will reach pareto optimal solutions by itself and government
regulation is not necessary. Buchanan and Stubblebine [11] and
Turvey [49] make even stronger statements. They show that with some
types of government intervention, such as effluent taxes, subsequent
bargaining by victims will disrupt the efficient allocation.
Intervention, if negotiation is possible, may actually make matters
worse.

Not all observers agree that government intervention in pollution
is unnecessary or harmful. Wellicz [54] argues that it is not clear that
the victim and polluter would bargain over marginal quantities. The
negotiation process is like a two person game and with strategic
maneuvering, it is not clear that the final outcome would necessarily
be pareto optimal. Further, as Dolbear [17] emphasizes, if the parties
are individuals, the income effects from negotiations could alter the
marginal conditions. The resulting arrangement may not be pareto optimal
in terms of the original distribution of income.

The critical argument for government intervention in air pollution control, however, is that there is not one victim but rather a multitude of victims. An optimal solution could only be reached if all the victims jointly negotiate with the polluting firm. Mishan [55], Baumol [7], and Baumol and Oates [9] all note the similarity between this joint decision and the model by Samuelson [45] of public goods. As with public goods, the private market cannot arrive at an efficient solution because only one quantity of the good is chosen for all (one level of pollution)[3], because there is no way to avoid free riders (people who benefit without paying), and because transactions costs are substantial. Negotiations by individual victims is suboptimal under these conditions and intervention is appropriate.[4] Though not universal, the need for government intervention in pollution control is at least widespread.

Of course, actual government intervention, even when needed, is not by definition beneficial. Pigou suggests that government should equilibrate private and social net product. The producers of pollution should incorporate the damages of their emissions into their production decisions. Subsequent authors have developed the distinction between costs to polluters and costs to victims. Kneese and Bower [26], incorporate these distinctions into a formal cost minimization model. For every level of output, the total costs to society of air pollution are minimized when the marginal cost to the polluter of an additional unit of emission just equals the marginal damages that emission has upon all victims.

The presentation of total cost minimization through marginal benefit and cost curves has several advantages (see Figure I-A). First, given certain assumptions, it correctly depicts the efficient solution.[5]

Figure I-A
Minimizing Costs with Marginal Analysis

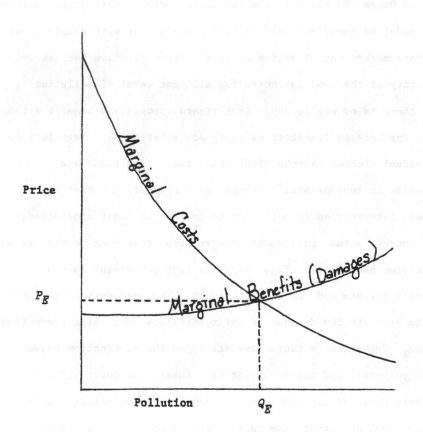

Total costs are minimized when marginal pollution damages
equal marginal pollution abatement costs. The efficient
price is P_E and the efficient quantity is Q_E.

Second, it easily demonstrates the implications of various pricing policies. Third, the distributional effects between victims and polluters of various policies are easily accessible through the analysis. For example, with a quantity control program, the emitter must pay an amount equal to the area under the marginal cost curve up to the efficient level. The victims suffer an amount equal to the area under the marginal benefit curve between the efficient level and no pollution at all.

A major drawback of this type of marginal analysis is that marginal benefit curves hide a number of important value judgments, especially judgments about the value of health or life. A large fraction of the benefits of air pollution control are non-marketable goods, items which are not normally traded. The two most important items are reduced population morbidity and mortality rates. In order to construct a marginal benefit curve, it is necessary to make judgments about the values of these goods. Presenting results through marginal cost and benefit curves obscures these critical value judgments and encourages experts to insert their own values into the analysis instead. As an alternative approach, the analysis presented here emphasizes the tradeoff between health losses and non-health costs (defined as the sum of the abatement costs and vegetation and material damages). Effective abatement devices are defined as methods which minimize total non-health costs for given levels of health effects (see Figure I-B). The collection of all these cost minimizing points represents the available tradeoff to society between the health effects and the non-health costs of air pollution. The most efficient point, where the reader may wish to be along that tradeoff, depends upon the value the reader places on

Figure I-B

The Tradeoffs Between Health and Non Health Costs

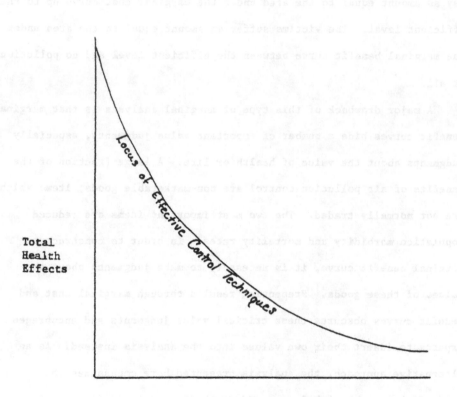

Total Health Effects

Locus of Effective Control Techniques

Total Non Health Costs

For every level of health effects, effective techniques
achieve the health goals with minimum costs. The slope
of this curve is the marginal expenditure for additional
health improvement. Techniques to the right of this
curve are less effective than methods along the curve.

health effects. Though it is inevitable that all analyses contain some hidden value judgments, it is imperative to leave as many of these value judgments to society as possible in order to preserve the integrity of economic analysis.

In addition to these efficiency issues, the control of air pollution is ripe with questions of equity. Atmospheric rights are almost as amorphous as the atmosphere itself. With firms and individuals vying for greater use of the atmosphere, who really owns the air is becoming an increasingly important issue. Air pollution control regulations, in addition to settling efficiency problems, also settle these equity issues. For instance, the type of control devices permitted will effect who must suffer the damages from the remaining pollution. Some devices favor people who live nearer to the plant and other techniques favor individuals who live further away. The regulatory tool itself will settle other equity issues. For instance, if the efficient solution is achieved through quantity controls, then the polluters are given the right to pollute up to the efficient amount. On the other hand, if the efficient solution is achieved through government subsidies for abatement, then polluters implicitly have the right to pollute as much as they want. Air pollution regulations are, therefore, going to settle important controversies about equity. Just as with questions of efficiency, careful analyses can help clarify these equity issues by identifying the alternatives and explaining their expected consequences.

Another important characteristic of air pollution management is that decisions often have to be made under great uncertainty. If this uncertainty is local or site by site, the uncertainty of the total outcome across many sites diminishes as the number of sites increase.

Across the country, there are so many individual sources of air
pollution, that this type of uncertainty is surely negligible on an
aggregate national basis. A large fraction of the benefits of air
pollution control, however, are local health effects. Since localities
are unlikely to be compensated for local aberrations in health effects,
each locality should be primarily concerned with the sources of air
pollution in its vicinity. Each locality, each individual, would
understandably be risk averse with respect to the local application of
even nationwide air pollution control policies. In contrast to this site
by site uncertainty, many types of uncertainty apply uniformly to all
sources. For instance, little is understood about the consequences of
exposures to most air pollutants.[6] Some abatement techniques may be
wasting substantial resources controlling relatively harmless pollutants
while permitting large quantities of harmful pollutants to pass over
large population areas. Since this mistake could be repeated across
many emission sites, the uncertainty does not diminish with increasing
numbers of sources. Control techniques which are robust, even in the
midst of this type of uncertainty, are clearly preferable.

CHAPTER I-B

REVIEW OF APPLIED AIR POLLUTION CONTROL MODELS

This study explores policies which effect the choice of air
pollution control techniques for given emission sources. Although one
is concerned about how these decisions might alter the costs of
production, how these changes in the cost of production will, in turn,
effect the final demand for products is beyond the scope of this study.[7]
A number of studies have explored various ways to control air pollution
problems. These studies not only explore and choose different pollution
control techniques, but they also define the air pollution problem in
quite different terms.

Both EPA [52] and Kohn [27] define the air pollution problem as
minimizing the cost of reducing aggregate emissions by a prespecified
amount across firms in one airshed. The specific purpose of the linear
programming models of both authors is to identify how much each firm
should reduce its emissions. Whether the aggregate amount of pollution
reduction is efficient is not addressed in this model. Another
simplification is that both models assume that an emission anywhere in
the airshed has the same effect (this is not supported by simple
meteorological evidence). Because the costs of abatement vary across
firms, the model demonstrates that it is cheaper for society to require
some sources to reduce emissions more than others. For instance, a
regulation that all sources reduce sulfur emissions by 80% is estimated
to cost $144/ton for steam electric plants, $550/ton for area sources,
$600/ton for petroleum, $277/ton for sulfuric acid, and less than $100/
ton for primary nonferrous smelters. The EPA and Kohn models clearly
demonstrate that uniform emission reductions across industries can be

more expensive than at least some schemes of differential emission reductions.

Atkinson and Lewis [4] define the air pollution control problem in terms of minimizing the cost of achieving an average level of ambient air concentrations. This definition of the air pollution problem recognizes the importance of the location of the emission source and general atmospheric conditions. Though this advance should yield substantial benefits, Atkinson and Lewis fail to capitalize upon their insights. Instead of a maximum ambient air standard in every location, Atkinson and Lewis use an average air quality constraint across the entire metropolitan area. By averaging across the entire metropolitan area, the model mistakenly treats the ambient air concentrations of pollutants equally regardless of their location in the metropolitan area. Because it is frequently easier to keep rural rather than urban air clean, the solutions to the model often entail high pollution concentrations in the city and low concentrations only in remote rural areas. If the bulk of the population lives in or near the central city, this policy will lead to inordinate population exposures and consequently greater damage.

Russell and Spofford [44] improve upon the Atkinson and Lewis formulation by imposing average ambient air standards on disaggregated areas within the metropolis. The Russell-Spofford model minimizes costs with respect to prespecified local maximum ambient air concentrations for each pollutant. This model does not determine an efficient level of ambient air concentrations—the model simply calculates the cheapest way to meet an exogenous air quality standard. An added feature of the Russell-Spofford model is that it examines all forms of waste: air,

water, and soil. By examining all wastes simultaneously, the authors
avoid treating wastes which are not of central concern as though they
have no social cost associated with them. Choi, Dee, and Requum [13]
and Ayres [5] chide other studies for ignoring all these social costs.
The criticism is well taken and one should indeed account for the social
costs of disposing of air pollution in water or the soil. It is not
clear, though, that one needs to build an enormous framework that simulates
each form of waste simultaneously. This holistic approach to the
environment has appealing qualities, but it makes any model complex and
cumbersome and may actually impede the overall analysis.

All the studies just discussed are concerned with air pollution
from stationary sources. Another more complex and decentralized air
pollution problem is mobile source air pollution. Automobile exhausts
are more difficult to study because it is harder to predict local
emissions and because less is known about the effect of control policies
on actual emissions. Ingram and Fauth [24] make an important
contribution to understanding mobile source pollution by sequencing an
urban transportation planning model, an emission submodel, and a
dispersion model. The transportation section generates trips between
zones and modal choices, the emission model converts these trips into
tons of effluents in each zone, and the diffusion model predicts ambient
air concentrations according to a modified Gaussian plume. The
researchers choose not to tackle the difficult nonlinearities of chemical
interactions in the atmosphere but they suggest extensions of their model
in this direction would be helpful. Though Ingram and Fauth do not
integrate their simulation into a formal maximization process, they do
compare the costs and benefits of selected policies. Because the model

has no damage function, benefits are measured in terms of reductions in carbon monoxide and an exposure index which is the product of the population times exposures above the ambient air standards.

One limitation of the models just discussed is that the desired level of ambient air concentrations is never analyzed itself. In order to examine the ambient air standards, it is important to quantify the final consequences of different levels of air quality. One way to associate levels of pollution with final effects is to use dose response functions. Of all the dose response functions for air pollutants, the effects of radioactive materials received the earliest and greatest attention. It is not surprising, then, that the first studies to include dose response curves are analyses of the nuclear energy industry. EPA [51] analyzes the entire light water nuclear fuel cycle including mining, transportation, processing, and the power reactor. The costs of additional water and air abatement techniques are compared with the benefits from each process. The benefits are calculated from a relatively sophisticated environmental model. Natural dispersion through water, the atmosphere, and food chains are traced in order to measure the total population exposure. This exposure is then translated by dose response curves into health effects which are the sum of deaths and cancerous illnesses (they, unfortunately, do not keep these effects separate). This study is one of the first pollution analyses to examine abatement choices in terms of the costs of achieving certain health effects.

North and Merkhofer [34] attempt to compare the costs and benefits of air pollution control of sulfur emissions from power plants. The model starts with the supply and demand for electricity, predicts emissions, simulates their dispersion, determines expected effects, calculates dollar damages, and then chooses optimal control strategies. Their environmental economic model, unfortunately, does not realize their ambitious research outline. Only a few illustrative abatement techniques are examined. The importance of site choice is shown with only two cases, an urban plant and a rural plant located 80 kilometers away. Dispersion is modelled in a simple geometric fashion which provides only crude estimates of ambient concentrations. The exposed population (New York) appears to be subject to uniform ambient air concentrations despite the non-uniform distribution which the diffusion model generates. Also, little was understood at the time of this study about the specific dose responses towards individual pollutants. North and Merkhofer make an attempt to include damages to soil, forests, and aesthetics but these effects are not integrated with the model's analysis of pollution dispersion.

Barringer, Judd, and North [6] expand the scope of the North and Merkhofer model by including the entire fuel cycle of coal and nuclear fuels. The expanded version, however, makes no improvements in the environmental model and consequently it is more illustrative than analytic. The framework of this analysis of the entire fuel cycle is relatively complete, however, and it could serve as the foundation for future work.

Although reducing air pollution emissions may be an effective way of improving air quality, careful selection of emission sites can also significantly improve the quality of air in selected locations such as dense metropolitan areas. Site selection, however, has rarely been studied as an air pollution abatement process. One of the only cost

benefit studies of power plant siting is the work by Niehaus, Cohen, and
Otway [53]. The authors take the expected accident risk of a nuclear
power plant from an AEC report [50] and examine how ultimate damages
would change as one moved a plant nearer or further from a hypothetical
population center. The population in the model is distributed in three
concentric circles surrounded by an infinite rural area. The densities
are homogeneous within each ring and decline with distance from the
center. Given the dose and population distribution, final "risk" is
calculated with a linear dose response function. These final risks are
measured in terms of both acute deaths and total man-rems exposure.[8]
Using a guideline value of $1000 per man-rem and the transmission costs
of remote siting, the authors calculate that it does not pay to move
plants out of the center city.

One should accept these negative results on remote siting with some
caution. First, the authors use the expected number of "risks" (health
effects) and ignore the added psychic costs of catastrophic accidents
where many people would die at once. Given that society has already
expressed a preference about catastrophic versus continual, random
accidental deaths (note jet versus automobile accidental deaths), it may
well be that the value of a single large event is greater than the
expected number of lives lost (as though they were small random events).
Second, the authors ignore the substitutions between emission control
costs and distance from the population. The AEC document which they rely
upon demonstrates maximum feasible safety equipment. With such equipment
on a plant, it is not surprising that the marginal benefits of moving it
away from a population center are low. That, however, does not mean that
society would not be better off with less safety equipment and plants

outside of central cities. Third, the authors report that the marginal benefits of moving a plant away from the central city are negative for the first 20 kilometers. Ambient concentrations from a source clearly decline dramatically over the first 20 kilometers and so the first few kilometers should have the greatest benefits.

Although the goal of understanding the benefits from air pollution control is generally accepted, how to evaluate these benefits is subject to considerable debate. The approach developed here is to quantify the physical effects of air pollution because it is easier to evaluate these final effects than an invisible intermediate product called air quality. An alternative approach is to evaluate society's revealed preference for higher air quality. Several researchers try to measure this preference for clean air by analyzing property value differentials between areas with and without clean air. These differences are assumed to reflect what people would be willing to pay for cleaner air. Ridker and Henning [67], Anderson and Crocker [2], Jaksch and Stoevenor [25], and several other authors have all regressed owner occupied housing values in single SMSA's on air pollution gradients and other housing characteristics. The researchers assume that, ceterus paribus, homes should have higher values if located in areas with better air quality. The regression technique searches for this pollution rent gradient after making adjustments for differences in other housing characteristics. The beauty of this approach is its use of market information to reveal what people are willing to pay for air quality. After adjusting their equations, most researchers are able to find "significant" negative coefficients for their pollution variables.

Despite the proliferation of housing value studies, they have such severe problems that they are almost worthless. Most of the property value studies use census tracts for observations. Because such data is an aggregate of many households and because it only provides a limited number of attributes, it is difficult to properly control for housing and neighborhood quality. Especially because these studies use only primary pollutants, lower air quality may be closely associated with other potential causes of lower property values which may be near industrial or central city locations. It is not at all clear whether these property value studies have been able to remove these unwanted influences from their analysis.

In addition to these severe data problems, Freeman [19] argues that the rent gradient can only provide the marginal willingness-to-pay of individuals given a particular spatial distribution of pollution. This information is not equivalent to what people would pay to have the entire distribution altered dramatically. Only a general equilibrium model could provide an assessment of what people are willing to pay for large changes in the level of pollution abatement. Unfortunately, as Polinsky and Shavell [39] demonstrate in their attempts to build such a model, realistic general equilibrium models are extremely hard to construct. Freeman [20] offers an additional model which utilizes cross sectional information from several cities. Assuming that all other factors across cities are controlled, one can estimate willingness-to-pay by comparing the rent differentials across cities. It is a difficult exercise to control for unwanted variations across cities, however, so that it is not clear whether such an analytic method would provide much of an improvement.

A final conceptual problem with the property value approach
surrounds the question of what people know about the effects of
pollution. Ayres [5] argues that real estate values probably reflect
experiential aspects of pollution such as soiling, smells, dirt, and
minor irritations. The most menacing air pollutants would be judged
mildly on these grounds. It is not clear that people even detect the
existence of all air pollution, much less comprehend its full consequences
upon their health and property. It is not clear what market actions
reveal about true, if informed, preferences. Pollution rent differentials
probably provide very little useful information about what an informed
society would want to spend to reduce air pollution. They do, at least,
suggest that the spatial allocation of households or at least the
distribution of housing prices will be effected by pollution control
strategies. Obtaining measures of the resultant utility loss or gain
from these changes, however, is not an easy task.

Another method of ascertaining pollution damage estimates is through
questionnaires. Ridker [40] performed one of the earliest surveys. He
interviewed a group of residents who were bombarded with soot following
an accident in a fossil fuel plant. He found a correlation between his
pollution index for the incident and house and auto cleaning costs by
residents. In an effort to obtain the psychic costs of the incident,
Ridker asked the residents what they would have paid to have avoided the
damaging effects altogether. Only 8 of the 85 respondents provide
willingness-to-pay estimates which are above their actual outlays.
Either the psychic costs of such incidents are low or the interview
method is a poor way of obtaining true estimates. Another survey done
in 1972 [21] finds that only 8% of national respondents are willing to

pay one hundred dollars or more "to improve natural surroundings."
Almost half are only willing to pay ten dollars or less per year.
Lawyer [28] finds the willingness-to-pay to have unnoticeable amounts of
air pollution in Morgantown, West Virginia by residents is near sixteen
dollars per person per year. Williams and Bunyard [55] find 66% of their
sample only want to spend five dollars per year to clean the air of
St. Louis (in 1963). These estimates of what people will pay for local
improvements may not be indicative of what they would pay for national
improvements. Nonetheless, it is remarkable how consistently low these
estimates are compared to the relatively high estimates of the property
value studies. Should these estimates of willingness-to-pay be
believed? Probably not. As Samuelson [45] notes in his seminal paper,
there are no incentives for a citizen to reveal his true preferences for
public goods. Secondly, as noted earlier, it is not clear that people
are informed of the consequences of current pollution levels. Answers
to questionnaires may be misleading sources of information about
society's preferences for clean air.

Most of the models reviewed in this analysis attempt to minimize
costs under various constraints. In order to solve these models, many
researchers have turned to linear programming. Although these models are
consequently easy to solve, there are several features of environmental
analysis which violate the assumptions of linear programming models. For
instance, many abatement methods have nonlinear scale effects.
Russell [43] notes that one electrostatic precipitator may remove 80% of
potential flyash, but a second device may only remove an additional 5%.
This type of nonlinearity is cumbersome to embody in a linear programming
framework. Environmental phenomenon also frequently are nonlinear.

Russell, Spofford, and Kelly [46] embed a sophisticated nonlinear aquatic dispersion model into their overall model which is solved through a series of iterations. The iterations end when the solution is arbitrarily close to the prespecified environmental standards. Unfortunately, the model is not easy to solve and the authors have not yet approached a full scale simulation. Nonetheless, nonlinear programming techniques offer an important opportunity to incorporate more realistic environmental models.

This study uses yet another solution algorithm, universal search. Every known abatement technique is examined as well as combinations of each technique. Though this approach is certainly less elegant, it permits a robust examination of nonlinearities and complex interactions.

Over the past several years, a substantial body of literature has addressed the problem of measuring and controlling the costs of air pollution control. Some of these analyses successfully respond to specific questions such as the effectiveness of uniform emission reductions. Only a few studies address the more basic problem of measuring both the costs and the benefits of pollution control. Despite their limitations, the environmental and economic models just reviewed serve as the foundation for the new model in this study. Hopefully, the analysis completed here will, in turn, serve as a building block for future generations of models.

Footnotes

[1] For a comprehensive review of environmental economics, see Fisher and Peterson [18] and for a review of the literature on externalities, see Mishan [30].

[2] See Pigou [38], p. 184 (underlining by editor).

[3] Since ambient concentrations decline with distance, it appears that victims have many choices of air quality. It remains true, though, that there is only one schedule of concentrations with distance and people may have different preferences for schedules.

[4] It should be noted that the government also finds the problem of allocating public goods difficult. They, too, have no way of determining true preferences for public goods. See Samuelson [45].

[5] The conditions for maximization are perfect competition and the necessary second order conditions. See Buchanan [10] for a discussion of the first assumption and Baumol and Bradford [7] for a discussion of the latter assumption.

[6] Though it is impossible to know exactly what is not known, some aspects of air pollution control are clearly not yet understood. The scientific evidence on the nature of air pollution dose response curves is especially weak. See the discussion of dose response functions for more details.

[7] There are a number of studies which examine the relationship between pollution levels and what should be produced. Leontief [29] and Ridker and Herzog [41] use input-output models and Griffin [22] and Chapman [12] use econometric models to address these relationships. None of the above studies, however, attempt to assess pollution control in terms of what could be gained from it.

[8] A man-rem is a measure of radiation dosage: one man being exposed to one rem of radiation. The rem is defined in terms of biological significance; one rem from any radiation source should have the same effect on a man.

Bibliography

1. Ackerman, B.; Rose-Ackerman, S.; Henderson, D.; and Sawyer. The Uncertain Search for Environmental Quality. NewYork: The Free Press, 1974.

2. Anderson, F.J., Jr. and Crocker, T.D. "Air Pollution and Residential Property Values." Urban Studies, VII (October, 1971), 171-180.

3. _____. "Air Pollution and Property Values: A Reply." Review of Economics and Statistics, LIV (November, 1972), 470-73.

4. Atkinson, S.E. and Lewis, D.H. "A Cost Effectiveness Analysis of Alternative Air Quality Control Strategies." Journal of Environmental Economic Management, I (November, 1974), 237-50.

5. Ayres, R.V. "Air Pollution in Cities." Natural Resources Journal, IX (January, 1969), 17.

6. Barringer, S.M.: Judd, B.R., and North, D.W. "The Economic and Social Costs of Coal and Nuclear Electric Generation: A Framework for Assessment and Illustrative Calculations for the Coal and Nuclear Fuel Cycle." Stanford Research Institute, Menlo Park, California, March, 1976.

7. Baumol, W.J. "On Taxation and The Control of Externalities." American Economic Review, LXII (June, 1972), 307-22.

8. _____ and Bradford, D.F. "Detrimental Externalities and Non Convexity of the Production Set." Economics, XXXIX (May, 1972), 160-76.

9. _____ and Oates, W.E. The Theory of Environmental Policy. Englewood Cliffs, New Jersey: Prentice Hall, 1975.

10. Buchanan, J.M. "External Diseconomies, Corrective Taxes, and Market Structure." American Economic Review, LIX (March, 1969), 174-77.

11. _____ and Stubblebine, W.C. "Externality." Economica, XXIX (November, 1962), 371-84.

12. Chapman, D. "Internalizing an Externality: A Sulfur Emission Tax and the Electric Utility Industry." Energy: Demand, Conservation, and Institutional Problems. Edited by M.S. Macrackis. Cambridge, Mass.: MIT Press, 1974.

13. Choi; Dee; and Requum. "Interdependence of Air, Water, and Soil Pollution Control Strategies." Journal of Water, Air, and Soil Pollution, (1975), 381-94.

14. Coase, R.H. "The Problem of Social Cost." Journal of Law and Economics, III (October, 1960), 1-44.

15. Crocker, T.D. Urban Air Pollution Damage Functions: Theory and Measurement. Environmental Protection Agency, Washington, D.C., January, 1971.

16. Deyak, T.A. and Smith, V.K. "Residential Property Values and Air Pollution: Some New Evidence." Quarterly Review of Economics and Business, XIV (Winter, 1974), 93-100.

17. Dolbear, F.T. "On the Theory of Optimum Externality." American Economic Review, LVII (March, 1967), 90-103.

18. Fisher, A.C. and Peterson, F.M. "The Environment in Economics: A Survey." Journal of Economic Literature, XIV (March, 1976), 1-33.

19. Freeman, A.M., III. "Air Pollution and Property Values: A Methodological Comment." Review of Economics and Statistics, LIII (November, 1971), 415-16.

20. _____. "On Estimating Air Pollution Control Benefits from Land Value Studies." Journal of Environmental Economic Management, I (May, 1974), 74-83.

21. Gallup, . "Public Opinion on Environment Sampled." Environmental Health Letter, XI (May, 1972).

22. Griffin, J.M. "An Econometric Evaluation of Sulfur Taxes." Journal of Political Economy, LXXXII (July/August, 1974), 669-88.

23. _____. "Recent Sulfur Tax Proposals: An Econometric Evaluation of Welfare Gains." Energy: Demand, Conservation, and Institutional Problems. Edited by M.S. Macrakis, Cambridge, Mass.: MIT Press, 1974.

24. Ingram, G.K. and Fauth, G.R. TASSIM: A Transportation and Air Shed Simulation Model. Report DOT-05-30099-5, Harvard University, Cambridge, Mass., May, 1974.

25. Jaksch, J.A. and Stoevenor, H.H. Effect of Air Pollution on Residential Property Values in Toledo, Oregon. Agricultural Experiment Station, Oregon State University, Corvallis, Oregon, Special Report 304, September, 1970.

26. Kneese, A.V. and Bower, B.T. Managing Water Quality: Economics, Technology, Institutions. Baltimore: Johns Hopkins Press for Resources for the Future, 1968.

27. Kohn, R.E. "Optimal Air Quality Standards." Econometrica, XXXIX (November, 1971), 983-95.

28. Lawyer, R.E. "An Air Pollution Public Opinion Survey for the City Morgantown, West Virginia." Unpublished Master's thesis, West Virginia University, Morgantown, West Virginia, 1966.

29. Leontief, W. "Environmental Repercussions and The Economic Structure: An Input-Output Approach." Review of Economics and Statistics, LII (August, 1970), 262-71.

30. Mishan, E.J. "Postwar Literature on Externalities: An Interpretative Essay." Journal of Economic Literature, IX (March, 1971), 1-28.

31. National Academy of Sciences. Air Quality and Stationary Source Emission Control. For the Committee on Public Works, U.S. Senate, Washington, D.C., Serial No. 94-4, March, 1975.

32. Nelson, J.P. The Effects of Mobile Source Air and Noise Pollution on Residential Property Values. University Park, Pa.: Institute for Research on Human Resources, 1975.

33. Niehaus, F.: Cohen,J.J.; and Otway, H.J. "The Cost Effectiveness of Remote Nuclear Siting." IIASA, Laxenburg, Austria, April, 1976.

34. North, D.W. and Merkhofer, M.W. "Analysis of Alternative Emission Control Strategies." Air Quality and Stationary Emission Control. National Academy of Sciences, March, 1975.

35. Nourse, H.O. "The Effect of Air Pollution on House Values." Land Economics, XLIII (May, 1967), 181-89.

36. Otway, H.J.; Lohrding, R.K.; and Battat, M.E. "A Risk Estimate for an Urban Sited Reactor." Nuclear Technology, XII (October, 1971), 173-84.

37. Peckham, B.W. "Air Pollution and Residential Property Values in Philadelphia." U.S. Department of Health, Education, and Welfare, National Air Pollution Control Administration, Raleigh, North Carolina, 1970.

38. Pigou, A.C. The Economics of Welfare. London: Macmillan & Co., Ltd., 1920.

39. Polinsky, A.M. and Shavell, S. "The Air Pollution and Property Value Debate." Review of Economics and Statistics, LVII (February, 1975), 100-104.

40. Ridker, R.G. Economic Costs of Air Pollution. New York: Frederick Praeger, 1967.

41. _____ and Henning, J.A. "The Determinants of Residential Property Values with Special Reference to Air Pollution." Review of Economics and Statistics, XLIX (May, 1967), 246-257.

42. Ridker, R.G. and Herzog, H.W. "The Economy, Resource Requirements, and Pollution Levels." and "The Model." Population, Resources, and the Environment, Edited by R.G. Ridker for U.S. Commission on Population Growth and the American Future, Washington, D.C., 1972.

43. Russell, C.S. "Models for Investigation of Industrial Response to Residual Management Actions." Swedish Journal of Economics, LXXIII (March, 1971), 134-56.

44. _____ and Spofford, W.O., Jr. "A Quantitative Framework for Residuals Management Decisions." Environmental Quality Analysis: Theory and Method in the Social Sciences. Edited by A.V. Kneese and B.T. Bower. Baltimore: John Hopkins Press, 1972.

45. Samuelson, P.A. "The Pure Theory of Public Expenditure." Review of Economics and Statistics, XXXVI (1954), 387-89.

46. Spofford, V.; Russell, C.S.; and Kelly, R. "Operational Problems in Large Scale Residuals Management Models." Economic Analysis of Environmental Problems. Edited by E.S. Mills. New York: Columbia University Press, 1975.

47. Spore, R.L. "Property Value Differentials as a Measure of the Economic Costs of Air Pollution." Ph.D. dissertation, Center for Air Environment Studies, Penn State University, University Park, Pa., CAES Publication Number 254-72, June, 1972.

48. Teitenburg, T.H. "Specific Taxes and Pollution Control: A General Equilibrium Analysis." Quarterly Journal of Economics, LXXXVII (November, 1973), 503-22.

49. Turvey, R. "On Differences Between Social Cost and Private Cost." Economica, XXX (August, 1963), 309-13.

50. United States Atomic Energy Commission. An Assessment of Accident Risks in Commercial Nuclear Power Plants. WASH-1400, Washington, D.C.

51. United States Environmental Protection Agency. "Environmental Analysis of the Uranium Fuel Cycle." EPA-520/9-73-003-C, Washington, D.C.: U.S. Printing Office, 1973.

52. United States Environmental Protection Agency. "Cost Effectiveness of a Uniform National Sulfur Emissions Tax." Research Triangle Park, N.C., 1974.

53. Weiand, K.F. "Property Values and the Demand for Clean Air: Cross Section Study for St. Louis." Presented at Committee on Urban Economics Conference, Chicago, Illinois, September 11-12, 1970.

54. Wellicz, S. "On External Diseconomies and the Government Assisted Invisible Hand." _Economica_, XXXI (November, 1964), 345-62.

55. Williams, J.D. and Bunyard, F.L. "Interstate Air Pollution Study: Opinion Surveys and Air Quality Statistical Relationships." U.S. DHEW, Public Health Service, Washington, D.C., 1966.

56. Zerbe, R.O. _The Economics of Air Pollution: A Cost-Benefit Approach_. Report to the Ontario Department of Public Health, Toronto, Ontario, Canada, 1969.

CHAPTER II

INTRODUCTION TO THE MODEL

The model used in this paper is designed to evaluate the costs, benefits, uncertainties, and distributional effects of air pollution abatement techniques. In order to shed some light on these issues, available scientific information is organized into an environmental simulation model. The purpose of the simulation model is to identify the final consequences of engaging in various abatement techniques.

The simulation model is composed of four logical steps which are shown in Figure II-A. The first step is to determine how much of each type of pollutant a given abatement technique will remove or add to the existing stream of effluents. This is largely an engineering problem and it is described in detail in Chapter III. The outputs of this step are the inputs of pollutants into the environment and the costs of implementing each abatement strategy. The next logical step is to determine where the pollutants go when they leave the plant and how they might change while travelling. This is largely a meteorological problem and is described in detail in Chapter IV. The output of this step is a description of the ambient concentrations of all pollutants at ground level for relevant locations. Utilizing data about the population distribution in all nearby areas, total exposures to the population can be calculated given the ground level concentrations of each pollutant. The specific application of this procedure is discussed briefly in Chapter VI. Calculating the final damages from the exposures is the purpose of the last step of the model. Several scientific disciplines are concerned with this problem including forestry, epidemiology, toxicology, and industrial chemistry. The final effects include excess mortality, morbidity, material losses, vegetation damage, visibility loss, and acid rain effects. Each of these are discussed in greater detail in Chapter V.

There are a number of basic assumptions in this model. One important assumption of the model is that it is possible to analyze a power plant or a single site independently of other emission sources. It is assumed that the effect of the power plant is the same regardless of the background level of air quality. There are at least two natural phenomenon which could violate this assumption: chemical interactions in the atmosphere and nonlinear dose response functions. Chemical interactions in the atmosphere may depend on the background level of air pollutants. For instance, in the presence of background hydrocarbons, the sulfur dioxide emitted from the power plant may change more rapidly into sulfate. The importance of such interactions is tested in the sensitivity analysis in the atmospheric modelling chapter. Nonlinear dose response functions could also create a connection between the marginal effects of a power plant and background levels of pollution. As explained in Chapter V, within the range of normal exposures, careful analysis is unable to detect nonlinear dose response curves. The evidence supports the assumption that the marginal dose effect of a harmful pollutant is independent of the background level.

Another assumption made in order to calculate the variance of the final outcomes is that each submodel is statistically independent of the other components of the model. It seems reasonable to assume that the errors in one component would be uncorrelated to the errors in another. Since the outcome is merely the product of the results of each submodel,

Figure II-A
Flow Chart of the Environmental Submodel

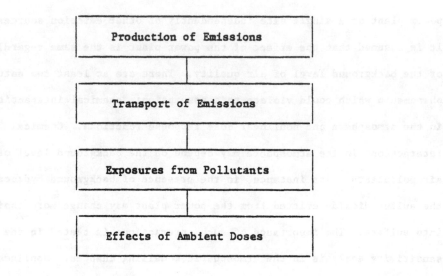

Production of Emissions

Transport of Emissions

Exposures from Pollutants

Effects of Ambient Doses

the variance of the outcome can be calculated from the following
equation:[2]

$$\text{Var} \left(E(I) * T(I,J) * D(J) \right) = \overline{E} \, (\text{Var} \, T) \, (\text{Var} \, D) +$$
$$\overline{T} \, (\text{Var} \, E) \, (\text{Var} \, D) + \overline{D} \, (\text{Var} \, E) \, (\text{Var} \, T) + \overline{E} \, \overline{T} \, (\text{Var} \, D) +$$
$$\overline{E} \, \overline{D} \, (\text{Var} \, T) + \overline{T} \, \overline{D} \, (\text{Var} \, E) + (\text{Var} \, E) \, (\text{Var} \, T) \, (\text{Var} \, D)$$

A third basic assumption of this modelling effort is that it is
possible to bring together evidence collected from a wide variety of
sources and experiences. There is always the possibility that what
appears to be a general truth in one context has no validity in another.
It is assumed that the data collected from each scientific discipline is
generally valid and is specifically applicable to the case study
considered here. Unfortunately, it is beyond the resources of this
study to test the validity of this last assumption.

CHAPTER III

AIR POLLUTION ABATEMENT ENGINNERING

There are a number of abatement strategies which may reduce the environmental damages associated with producing electricity from a coal fired power plant. Each strategy focuses upon a different method of reducing damages. Some techniques employ relatively "clean" fuels. Some alter the method of combustion or treat the flue gases which result from combustion. Finally, others disperse the flue gases in order to reduce total damages. Table III-1 lists all the abatement techniques considered in the analysis. In addition to these individual methods, there are a multitude of mixed strategies which involve combinations of each technique. Since the effects of these combinations are not linearly related to the sum of the individual parts, several combinations are examined separately.

Each abatement technique is described briefly in this chapter. Each description is followed with estimates of the expected cost of using each abatement method. The expected effect of the abatement technique is also presented in terms of how the method will alter the immediate output of the plant or similarly, the input from the plant into the environment.

Although some of the abatement techniques listed in Table III-1 began operation several decades ago, the immediate consequences and the costs of each method are not known with great certainty even today. It is quite possible that some of the ramifications of these techniques are not even suspected. Even if suspected, many effects have not yet been adequately measured. Also, site and operating conditions vary significantly across utilities and these variations drastically alter

Table III-1

The Air Pollution Abatement Techniques Examined in This Study

Sulfur Abatement
Coal Beneficiation
Flue Gas Desulfurization
Low Sulfur Coals
Fluidized Bed Combustion

Particulate Abatement
Electrostatic Precipitators (ESP)
Venturi Scrubbers
Air Bag Filters

Nitrogen Oxide Abatement
Staged Combustion
Flue Gas Recirculation
Flue Gas Scrubbers

Carbon Monoxide Abatement
Afterburners
Flue Gas Recirculation

Broad Based Pollution Control Methods
Efficient Use of Fuel
Tall Stacks
Reheating Flue Gas
Siting

costs and performance. Further, the cost and effectiveness of some
abatement methods over equipment lifetimes are often misunderstood
because they are new products, new designs of old products, or they are
being operated under new conditions. For all these reasons, one cannot
be certain of the costs or the outcomes of various abatement strategies.
In order to preserve a sense of this uncertainty, measures of the possible
variation around each estimate are also presented.

This chapter is organized around methods to control specific
pollutants. Techniques which specifically control sulfur oxides are
discussed first. This is followed by presentations of particulate,
nitrogen dioxide, and carbon monoxide control. The chapter concludes
with a review of methods which are not specific to any one pollutant but
which reduce the damages of air pollution in general.

Sulfur Control

Sulfur emissions from coal-fired power plants originate in the fuel
as pyrite (Fe_2S) or as a part of organic compounds. This distinction is
important because some abatement techniques are only able to remove
pyritic sulfur. Upon combustion, both sources of sulfur are oxidized
into sulfur dioxide. The sulfur dioxide is removed from the boiler along
with all the other flue gases and ejected into the atmosphere. During
combustion and during transit through the emission systems, some of the
sulfur dioxide can be further oxidized into sulfur trioxide. Sulfur
trioxide, in turn, can be absorbed by water droplets into sulfuric acid
mist or onto particles as sulfate. These transformations are relatively
slow in an uncontrolled power plant and, on average, 98% of sulfur

emissions are sulfur dioxide.

Coal beneficiation is a process whereby coal is cleaned before it is shipped away from the mine mouth. Physical coal cleaning generally involves crushing the coal and then floating the coal upon a liquid. The heavier pyrite particles tend to fall to the bottom leaving a coal with up to 40% less sulfur. How much sulfur is removed depends on the level of beneficiation and the percent of pyrite in the coal. Considerable ash is also removed in this process so that the cleaned coal has a higher BTU content per ton and produces fewer particulate emissions. This process not only reduces air pollution emissions, but it also reduces coal transportation costs (per BTU) and ash removal costs. These savings help reduce the net cost of coal beneficiation. Though beneficiation is a relatively simple process, a full scale demonstration of the technique for purposes of sulfur control has yet to be implemented.[1]

Engineers have estimated both the costs and effects of coal beneficiation from model plants and preliminary designs. The annualized cost lies between $1.23 and $2.16/ton of coal for Level 3 cleaning and between $2.90 and $4.84/ton of coal for more complete Level 4 cleaning. One of the advantages of cleaned coal is that it has a higher BTU content than run-of-mine coal. In order to include this benefit, the cost of coal cleaning can be expressed in terms of mills/kwhr (see Table III-2). Level 3 cleaning costs about 1 mill/kwhr while Level 4 cleaning costs about 2 mills/kwhr. As can be seen in Table III-3, Level 3 cleaning removes at least one-third and Level 4 cleaning removes at least two-fifths of the sulfur in most coals. Because the process only removes pyritic sulfur, the effectiveness of beneficiation depends heavily upon the percent of pyritic sulfur in each coal. Beneficiation is thus more

37.

Table III-2

The Cost of Coal Beneficiation
of Mid-Western and Appalachian Coals[a]

	Level 3	Level 4
Capital Costs ($/tons/yr)		
Large Plant	3.73	8.40
Small Plant	4.35	9.00
Operating Cost ($/ton)	.30-.60	.80-1.60
Annualized Cost (mills/kwhr)		
Large Plant	.73-1.05	2.00-2.70
Small Plant	.80-1.20	2.20-3.20

[a] Source: U.S. Dept. of Commerce [25].

Assumptions: 500 MW power plant, 6500 hrs/yr, 37% thermal efficiency.
Large scale coal treatment plant - 3,000,000 tons/yr.
Small scale coal treatment plant - 1,000,000 tons/yr.

effective in Northern Appalachian and Mid-Western (Western) coal than in the other Eastern coals. Another advantage of cleaned coal is the reduction in total ash wastes. Table III-3 shows how much ash is removed by each level process. Since ash wastes are presumably more easily disposed near remote mine sites than near urbanized power plants, coal beneficiation may reduce waste disposal costs between .2 and 1.5 mills/kwhr. If these savings are near the higher end of the range of estimates, the reduced disposal costs will pay for Level 3 cleaning.

In addition to physical coal cleaning, there are a number of chemical coal cleaning techniques which are being developed.[2] Table III-4 contains a list of these methods as well as their estimated cost per ton and the percent of sulfur which they are expected to remove. The chemical processes entertain the possibility of removing greater percentages of the pyritic sulfur and possibly some of the organic sulfur. These improvements come at substantial increases in cost. Most of the chemical processes will cost between five and ten times as much as physical coal cleaning. There is also greater uncertainty about the consequences of chemical coal cleaning since it is a relatively undeveloped technology.

The choice of fuels has a critical impact upon potential emissions of sulfur and upon sulfur removal systems. The sulfur content in coal can range from .5 to 6.0 percent depending upon where it is mined (see Table III-3). Most of the low sulfur coal in the country is in the West, low in BTU content, and high in ash. The Southern Appalachian area contains significant quantities of low sulfur, low ash, but this coal is primarily used for metallurgic processes. The Northern Appalachian and Mid-Western regions currently provide most of the coal used in the United States. The sulfur in these coals averages about 3 and 3.7 percent,

Table III-3

Enhancement of Coal Quality by Beneficiation[a]

	As Mined	Level 3 Beneficiation	Level 4 Beneficiation
Northern Appalachian			
lbs Ash/10^6 BTU	30.2	10.9	8.3
lbs SO_2/10^6 BTU	5.08	2.90	2.32
Southern Appalachian			
lbs Ash/10^6 BTU	25.4	5.8	5.2
lbs SO_2/10^6 BTU	1.61	1.10	1.08
Mid-Western (Eastern Block)			
lbs Ash/10^6 BTU	29.3	10.4	7.7
lbs SO_2/10^6 BTU	6.52	4.06	3.57
Mid-Western (Western Block)			
lbs Ash/10^6 BTU	31.1	9.9	8.4
lbs SO_2/10^6 BTU	6.41	3.29	3.0

[a] Source: U.S. Dept. of Commerce [25], Table 4, p.26.

Estimates are for average bituminous coal from each region.

Table III-4

The Cost and Sulfur Removal Ability

of Chemical Coal Cleaning[a]

Chemical Coal Cleaning Process	Annualized Cost ($ per ton of coal)	Percent of Sulfur Removed	
		Pyrite	Organic
Meyers	14–28	95	–
Hydrothermal	22–30	99	24–72
Hazen	18–25	80	–
KVP	25–35	99	13
LOL	24–49	95	–
BOM/ERDA	24–44	99	15
Physical Coal Cleaning Process	2–5	35–70	–

[a] Source: Hall [10], Table 24.

respectively, while the ash content is about 17 percent for both sources. The cost of using low sulfur coal involves four additional factors: BTU content, ash content, transportation costs, and regional mining impacts. Even taking into account the lower BTU content of Western coals (about 9000 BTU/lb instead of about 11,000 BTU/lb), Western coal emits only about one-fifth the sulfur that typical Eastern and Mid-Western coals produce. This benefit is partially offset by the increase in ash of Western coals. Transportation costs also play an important role in the cost of low sulfur Western coal. The further the power plant from the Western coal fields, the greater the premium for Western coal over high sulfur Eastern coal. For New England, in any case, low sulfur Western coal is expected to cost between 4 and 7 mills/kwhr more than high sulfur Eastern coal.

The transformation of coal into either gas or oil is another technique which allows nearly complete removal of the sulfur content of fuel. Though liquification of coal may be practical in the future, commercial demonstrations of this process will probably not occur before 1985, and it is uncertain whether the process will ever be economically competitive.[3] Coal gasification, on the other hand, has been under study over the last two decades and the processes necessary for full scale operation are understood. Nonetheless, there is still considerable uncertainty about the costs and environmental effects of a full-sized coal gasification plant. Current estimates suggest such a plant could produce low sulfur gas at a cost between $3.7 and $8./$10^6$ BTU. This is rather expensive when compared to run-of-mine coal at $.64/$10^6$ BTU or Level 4 beneficiated coal at $.80/$10^6$ BTU. Even if the purity of the final fuel is high and the environmental damage of the entire process is low, coal gasification is so much more expensive than the other techniques described here, it is unlikely that gasified coal would ever be used in large furnaces.

Another set of sulfur control methods focus on the conditions of combustion. Fluidized bed combustion may lower the amount of sulfur emitted in the flue gas to between 1% and 10% of uncontrolled emissions. This is achieved by crushing the coal and burning it in a bed of limestone or dolomite particles which adsorb the sulfur dioxide as it is released forming calcium sulfite or sulfate. These waste products can then be removed directly from the boiler.

The effectiveness of fluidized bed combustion depends upon several factors including the pressure in the boiler. At present, pressurized fluidized bed combustion is still largely in a developmental stage. As a result, the costs of these techniques are somewhat uncertain (see Table III-5).

Though most changes in boiler combustion are unlikely to alter the total mass of sulfur emitted into the flue gas, some methods may alter the proportions of sulfite (SO_3), sulfuric acid (H_2SO_4), or sulfate (XSO_4) in the flue gas.[4] Because these sulfur products appear to be far more hazardous than most emissions, even increases as small as a few percent can raise total damages significantly. The control of sulfur oxidation is, therefore, an important factor in air pollution management. Combustion parameters such as boiler residence time, combustion temperature, percent excess air, and pulverized coal might be important control techniques because of their potential effect on sulfur oxidation and small particle adsorption rates.

Table III-5

The Cost of Several Sulfur Removal Systems[a]

(figures in parenthesis are estimates of standard deviations)

	Capital Cost ($/kilowatt)		Operating Cost (mills/kwhr)		Annualized Cost (mills/kwhr)[b]	
Flue Gas Desulfurization						
Lime	61	(9)	3.0	(.5)	4.9	(.5)
Limestone	68	(9)	2.8	(.5)	4.9	(.5)
Magnesium Oxide	72	(9)	3.5	(.8)	5.7	(.8)
Wellman Lord	84	(9)	4.0	(.8)	6.6	(.8)
Catalytic Oxidation	90	(9)	4.4	(.8)	7.2	(.8)
Double Alkali	70	(9)	3.0	(.8)	5.1	(.8)
Fluidized Bed Combustion	75	(20)	4.0	(1.5)	6.0	(2.0)

[a] Source: Engdahl [7], Noll and Davis [19], Devitt [5], and U.S. EPA [27].

Assumptions: 500 MW plant, 6500 hrs/yr, onsite disposal of sludge, 3.5% sulfur.

[b] Includes charges for capital and overhead equal to 20% of capital costs.

A final method of sulfur control is to treat the flue gases. In 1976, thirty power plants in the United States were equipped with flue gas desulfurization (FGD). Removing sulfur from the flue gas of power plants is difficult because the sulfur diffuses with large volumes of waste gases. In order to remove the sulfur gases, it is necessary to concentrate the sulfur. Almost all FGD systems concentrate the sulfur by spraying the exit gases with chemical compounds which change the sulfur dioxide into sulfite or sulfate. The chemical composition of the sludge, the quantity of sludge, and the degree to which the sludge is treated varies across each FGD system.

Limestone and lime scrubbing are both throwaway processes. No useful byproducts are generated from the sludge. These processes differ in that one uses limestone and the other lime (calcined limestone) to scrub the flue gas. Both processes produce calcium sulfite and sulfate for disposal. These two processes are currently the most widely used FGD systems by utilities and consequently, more is known about their operation and costs than other systems. Though both of these systems have now been operated successfully on a commercial scale, problems with scaling plague many of the systems. The oxidation of calcium sulfite into calcium sulfate (gypsum) results in serious operating failures. By altering the pH levels, continuously monitoring sulfite concentrations, adding magnesium oxide, and cleaning the scrubbers frequently, engineers are able to keep these scrubbers in continuous operation.[5] The importance of good operation of these scrubbers should not be overlooked. Simply installing scrubbers will not result in substantial reductions of emissions over several years. It is imperative that these scrubbers be operated within relatively narrow specifications in order for their effectiveness to be realized.

The remaining FGD processes all treat the sludge in order to conserve the solvent and to produce marketable byproducts from the sulfur (sulfuric acid or elemental sulfur). The Magnesia process uses magnesium oxide and sulfur salts in order to remove the sulfur dioxide. The magnesium sulfite is then processed to make sulfuric acid and more magnesium oxide. The Sodium process (Wellman Lord) washes the flue gas with water droplets and sodium salts. Sodium sulfate is then purged from the system and elemental sulfur is produced from the sludge. Catalytic oxidation involves converting the sulfur dioxide to sulfur trioxide and then combining it with water vapor to form sulfuric acid mist which is removed from the flue gas. All three of these regenerable systems have the advantage that they produce marketable byproducts and reduce undesirable quantities of sludge.

It is difficult to obtain consistent figures on the cost of FGD systems. A comparison of six references results in a wide range of estimates of both capital and operating costs for each FGD system. These differences are partially due to the uncertainty about costs, but they are also due to the particular assumptions each analysis has made about operations. For instance, costs are often measured in current dollars so that estimates in different years are not consistent with each other. In order to avoid this problem, all figures in this paper have been adjusted to 1978 dollars.[6] The cost of FGD systems also vary with the size of the plant and the percent of sulfur in the fuel. Larger plants have lower per unit costs and more sulfur in the fuel adds to per unit costs (see Table III-6). Whether a technique is new or proven may also effect the long run costs because innovations often decrease the costs of processes in use for a long time.

Table III-6

The Effect of Plant Size, Percent Sulfur Coal, Backup Systems, and Retrofitting upon the Cost of Flue Gas Desulfurization[a]

Incremental Effect Upon:

Altered Factor	Capital Cost ($/kilowatt)	Operating Cost (mills/kwhr)	Annualized Cost (mills/kwhr)
200 MW	+20.0	+.8	+1.4
1000MW	-17.0	-.7	-1.2
2% Sulfur	- 7.0	-.4	- .6
5% Sulfur	+ 6.0	+.3	+ .5
Back Systems	+14.0	+.6	+1.0
Retrofit	+ 9.0	+.5	+ .8

[a] Source: U.S. EPA [27].

Base Case: New 500 MW plant, 3.5% sulfur coal, limestone scrubber.

Six FGD systems are compared in Table III-5 using similar assumptions.
for each case. Lime, limestone, and double alkali systems appear to be
less expensive than the other systems. The degree of uncertainty in
these estimates, however, is large and it is conceivable that any of the
systems might actually be the most effective in certain circumstances.
Table III-6 demonstrates the sensitivity of these estimates of costs to
different assumptions about original operating conditions. High sulfur
fuels, in addition to raising costs, also decrease reliability. It is
not clear that scrubbers can consistently clean high sulfur fuels.

One of the more serious drawbacks of nonregenerable systems is the
awesome quantity of sludge which they generate. A lime scrubber on a
1000 MW plant produces 500,000 tons of sulfur sludge each year (this
does not include the 600,000 tons of ash from normal combustion). The
lime/limestone sludge is composed of calcium sulfate and other acidic
impurities. The sludge appears to be solid but, if disturbed, it will
turn to liquid. In this state, sludge ponds become man-made quicksand
pits. Another difficulty with the sludge is that it is highly acidic
which leads to leaching problems. There are two ways to handle the
instability of the sludge material. One can combine the ash and the
sludge in open pits with good drainage or one can oxidize the sludge and
form an impure gypsum which is a good landfill material. The leaching
problem can be addressed by lining the bottom of the pit with clay,
plastic, or rubber. Finally, one needs a location to store large quantities
of sludge and possibly transportation to that location. All these
factors combine to make sludge disposal an expensive enterprise. Sludge
disposal, for a throwaway FGD system, can add between $6 and $20 per
kilowatt of capital costs and between .9 and 2. mills/kwhr of operating

costs. The regenerable systems, of course, have much lower disposal
costs because they produce substantially less waste. Sludge disposal
costs can vary significantly across sites because of variations in types
of fuels, land prices, and soil types. In turn, these costs can account
for significant variations in the total costs of FGD systems across the
country.

There are several operating parameters which can alter both the
costs and the effectiveness of FGD devices. One option is to treat only
a fraction of the flue gas from the combustion chambers. For instance,
one could treat only one-half the flue gas, lower the size of scrubbers
needed, the stack reheat costs, and the percent of sulfur removed. The
type of coal one chooses to burn will also effect scrubber performance.
Chlorine in coal is known to increase the scaling in lime and limestone
scrubbers. Other trace elements, BTU content, and the moisture content
of the fuel may also effect operating conditions. The operation of the
scrubber, the slurry flow rate, magnesium added, and the pH, also effect
the percent of sulfur removed. Another important operating variable is
increased reliability. By purchasing backup equipment, it is possible
to keep an FGD system operating longer. Additional scrubber units, more
ductwork, or duplicate backup systems increase the overall system
availability. They also increase costs. An estimate of the cost of
increasing reliability is provided in Table III-6. It should be clear
from this discussion that it is a simplification to speak of an FGD
system as a fixed process with one level of output and one set of costs.
The operation of every FGD system can be modified substantially to meet
specific needs.

A possible drawback of FGD systems lies in their internal conversion of sulfur dioxide into more dangerous sulfuric acid mist or sulfate. If these more harmful compounds are permitted to reach the atmosphere, they would significantly reduce the overall effectiveness of the FGD systems. Exit gases from scrubbers, unfortunately, are rarely tested for their percentage of sulfate, sulfuric acid, or sulfur trioxide. Properly operating scrubbers may reduce these harmful emissions but poorly operated devices may increase them.[7] The actual effectiveness of scrubbers under a variety of commercial conditions is still quite uncertain. It is certain that continual maintenance and careful operation of the scrubber is a critical component of scrubber effectiveness.

Particulate Control

There are a number of ways to control the potential quantity of particulates emitted from a coal fired boiler. One effective method is to use high temperatures and sufficient oxygen for more complete combustion. Another method is to add certain catalysts to the fuel. Fuel additives, unfortunately, often result in potentially toxic emissions so that their use is somewhat restricted. The major thrust to control particulates has been to remove them from the flue gas as it leaves the boiler. There are three major processes which clean flue gas: scrubbers, electrostatic precipitators, and air bag filters. The nature, costs, and effectiveness of each of these techniques is described, in turn, in greater detail.

Though regulatory agencies frequently treat particulates as

homogeneous quantities, a typical sample of flue gas particles may contain over thirty compounds. Table III-10 contains a list of some elements found in the flue gas of coal. In addition to the list of elements in the flue gas, the chemical composition can also vary across samples.[8] The size of the particles is also known to be important. Particles in the .1 to 1 micron range are more respirable and are known to be more harmful than either smaller or larger particles. Eliminating the larger particles from the flue gas may substantially reduce the mass of compounds emitted but it is not clear how much it reduces the potential damage of the waste system.

Although most scrubbers are designed to remove sulfur dioxide, Venturi scrubbers are designed to remove particulates. These high energy scrubbers can remove nearly 99% of the particulate mass. Scrubbers, unfortunately, have difficulty removing small particles which reduces their overall effectiveness. The cost and effectiveness of Venturi scrubbers is illustrated in Table III-7.

Electrostatic precipitators (ESP) attempt to take advantage of the ability of particles to hold a charge. After being charged, the particles are attracted to oppositely charged plates which they eventually coat. The particles are then neutralized and physically knocked off the plates and fall into hoppers below. There are several known designs of ESP's for either high or low temperature flue gas.[9] Successful operation of an ESP requires a careful balance between knocking the plates too lightly and knocking too vigorously. If one shakes the plates too gently, the particles will not fall off the plates and they will eventually lower the ability of the plate to attract new particles. If one knocks too violently, the particles are reentrained into the flue gas. Reentrainment

Table III-7

The Cost and Collection Ability of Particulate Removal Systems[a]

System	Collection Ability (%)	Adjusted[b] Collection Ability	Operating Cost (mills/kwhr)	Capital Cost ($/kw)	Annualized Cost (mills/kwhr)
Electrostatic Precipitators					
Medium Energy	97	90	.8	22	1.4 (.5)
High Energy	99	93	1.2	38	2.3 (.5)
Very High Energy	99.5	98	2.0	56	3.7 (.8)
Fabric Filter					
Air-to-Cloth	14	98	85	1.4	13	1.8 (1.0)
Ratios:	5.4	99	93	1.6	18	2.2 (1.0)
	2	99.9	98	1.9	35	2.9 (1.5)
Cyclone	80	40	1.4	18	2.0 (.5)
Venturi Scrubber					
Low Energy	95	80	1.8	35	2.8 (.6)
Medium Energy	99	87	2.4	41	2.7 (.8)

a Source: U.S. EPA [27], Noll and Davis [19], and Anonymous [4].

Assumption: 500 MW plant, 6500 hrs/yr.

b Adjustment made for collection efficiency of small particles assuming that one-half of all damages are made by small particles.

causes between 30 and 60% of emissions in high efficiency electrostatic
precipitators according to GCA [9]. High resistivity in the fly ash is
another potential problem especially with low temperature ESP's. High
resistivity tends to neutralize the charges on the plates. Sulfur
trioxide can counterbalance this problem because it neutralizes the fly
ash (see Table III-9). Sulfur trioxide, however, may be more harmful
than particulates if it is emitted. Whether it is desirable to have
sulfur trioxide in the flue gas depends upon how much is actually
emitted. The cost of ESP's are illustrated in Table III-7.

A third technique of removing particulates from flue gas is by
passing the gases through air bag filters which trap the particles. The
filters can then be removed, cleaned, and returned. Recent advances in
filter material now permit hot flue gas to pass directly through large
bags of filter material. The ability to treat hot gas is an important
innovation because it is expensive to cool, treat, and reheat the final
exit gases. The major advantage of air bag filters is their ability to
capture small deadly-sized particles. One of their drawbacks is the
questionable life-time of the air bags themselves. Additional breakthroughs
in filter materials and better operations promise to correct this problem.

There are a number of operating parameters for air bag filters which
can alter both their costs and effectiveness. Better filter material,
dolomite in the flue gas, lower air-to-cloth ratios, and higher pressure
drops can all be used to increase collection efficiencies. Table III-7
and Table III-8 illustrate the overall cost and effectiveness of several
air-to-cloth ratios. Note the rapidly increasing marginal cost as one
increases the ratio. Note that estimates of the costs of air bag filters
are highly variable; they fluctuate systematically with the estimated

Table III-8

The Ability of Particulate Removal Devices
to Collect Particles of Different Sizes[a]

	Particle Size				Total Mass
	.1-.5 µ	.6-1.0 µ	1.1-2.0 µ	2.1-10.0 µ	
Air Bag Filters[b]					
* 2/1 Air-to- Cloth Ratio	99	98	99.5	99.9	99.9
* 5.4/1 Air-to- Cloth Ratio	95	90	96	99.3	99
* 14.1/1 Air-to- Cloth Ratio	80	80	93	98	98
Electroatatic Precipitators[c]					
High Energy	95	97	99	99.5	99.5
Medium Energy	80	85	90	95	95
Cyclone[c]	1	5	10	50	50
Venturi Scrubber[c]					
Medium Energy	50	95	99	99.8	99.8

[a] Collection ability is measured in terms of percent of mass removed.

[b] These figures are extrapolated from Power Generation [19] Drehmel [6].

[c] These figure are extrapolated from Shannon, Gorman and Park [24], Figure 11, p.73.

life-times of the air bags.

As mentioned earlier, the effectiveness of particulate removal
systems depends upon their ability to remove small particles and
possibly certain compounds. Table III-8 compares the collection
efficiencies of a scrubber, an electrostatic precipitator, and an air
bag filter. Only the air bag filter and high energy electrostatic
precipitator can remove substantial quantities of small particles.
An adjusted collection efficiency is presented in Table III-7 to
incorporate the ability of each technology to remove small particles.
The adjusted collection efficiency simply averages the overall
collection efficiency with the specific collection efficiency for the
.1 to 1 micron sized particles. Venturi scrubbers fall from a 99%
overall efficiency rating to an adjusted efficiency rating of 87% with
these adjustments while fabric filters fall from 99% to just 98%.
Table III-10 compares the flue gas from a scrubber and an electrostatic
precipitator to determine which elements are removed by each. The
discrepancies between the technologies are often due to the typical size
of the elements. In general, more toxic elements appear to be smaller.

The cost of particulate removal systems may vary with each
application. For instance, the more sulfur in the fuel, the easier it
is to operate an ESP system (see Table III-9). There are also returns
to scale with respect to the size of the power plant. The per unit costs
of installing particulate removal systems rises dramatically with smaller
power plants (see Table III-9).

Table III-9

The Effect of Plant Size and
Sulfur Content of Fuel on Particulate Removal Costs[a]

	Total Operating Costs (mills/kwhr)	Total Capital Costs ($/kilowatt)	Total Annualized Cost (mills/kwhr)
100 MW Plant			
* Fabric Filter			
5.4/1 Air-to-Cloth Ratio	1.9	21	2.5
* Electrostatic			
Precipitator (high)	2.1	67	4.1
* Venturi Scrubber	4.0	70	6.1
(medium)			
1000 MW Plant			
* Fabric Filter			
5.4/1 Air-to-Cloth Ratio	1.5	16	2.0
* Electrostatic			
Precipitator (high)	1.0	31	2.0
* Venturi Scrubber			
(medium)	2.4	40	3.6
.8% Sulfur Coal			
* Fabric Filter			
5.4/1 Air-to-Cloth Ratio	1.8	20	2.4
* Electrostatic			
Precipitator (high)	2.4	90	5.1
* Venturi Scrubber			
(medium)	2.5	41	3.8

[a] Source: U.S. EPA [27].

Table III-10

The Ability of Particulate Removal Devices
to Collect Different Elements[a]

Element	Venturi Scrubber	Electrostatic Precipitator
Aluminum	99.7	99.3
Antimony	99.4	96.1
Arsenic	92.5	99.9
Barium	99.1	99.9
Beryllium	99.3	98.0
Boron	94.1	95.3
Cadmium	93.0	96.2
Calcium	99.1	99.2
Chlorine	25.0	19.8
Chromium	90.1	87.2
Cobalt	97.4	98.5
Copper	99.3	99.2
Flourine	98.0	92.4
Iron	99.3	99.2
Lead	98.1	99.5
Magnesium	98.8	99.2
Manganese	99.6	98.8
Mercury	13.2	2.1
Molybdenum	56.8	80.6
Nickel	95.9	87.8
Selenium	97.8	72.3
Silver	95.3	98.7
Sulfur	37.8	12.2
Titanium	99.7	99.4
Uranium	98.0	98.5
Vanadium	97.5	97.6
Zinc	97.5	97.4
Total	99.7	99.3

[a] Source: U.S. EPA [27]. Collective ability is measured in percent of
mass removed.

Nitrogen Oxide Control

Nitrogen oxides are formed when oxygen and nitrogen atoms combine in the presence of high temperatures. The source of the nitrogen appears to be from both the fuels and the atmosphere. One way to control the formation of these gases is to reduce the amount of time nitrogen is exposed to both high temperatures and excess oxygen. Combustion modification techniques all operate on this basis. Another control method is to treat the flue gases. Reduction, scrubbing, and catalytic decomposition equipment potentially can remove large percentages of nitric oxides from the flue gas.

Staged combustion is a two step firing technique. In the first step, the fuel is fired at high temperatures with little excess oxygen. In the second step, the gases are cooled and excess oxygen is added to complete burning. If properly operated, staged combustion can reduce nitric oxide emissions by 35% in coal fired boilers. Another combustion modification technique is to recirculate some of the flue gas. Recirculation lowers the temperature in the boiler and reduces emissions by 10%. Recirculation, however, may increase emissions of particulates and carbon monoxide, lower thermal efficiency, and increase corrosion.

There are only a few demonstration plants which remove nitrogen from flue gas and none of these have been used on power plants. An ammonia reduction and scrubbing plant has been built in the United States. This plant has been able to remove up to 90% of the nitric oxide from exit gases but, unfortunately, it also emits large quantities of unprocessed ammonia. Higher collection efficiencies and better performances can reasonably be expected from future plants but the costs of these processes

is expected to remain high. Table III-11 displays the relative costs
and removal efficiencies for each NOX removal system. The figures in
Table III-11 are not well understood as evidenced by the high degree of
uncertainty around them.

Carbon Monoxide Control

Carbon monoxide, like nitrogen oxide, can be controlled through
combustion modification techniques. In contrast to the nitrogen oxides,
carbon monoxide is reduced with high excess air and high temperatures.
This reduction can be achieved by either recirculating the flue gas for
additional combustion or by installing afterburners which convert the
carbon monoxide into relatively harmless carbon dioxide. Recycling can
lower the thermal efficiency of the boiler and afterburners require large
quantities of fuel, so both are expensive to operate.

The carbon monoxide emissions of large utility boilers are generally
not dangerous because large plants produce relatively small quantities.
Electric generating stations are not a major source of society's carbon
monoxide production. Because the danger of uncontrolled emissions is
small, carbon monoxide control techniques for utility-sized boilers have
rarely been studied. The accuracy of the estimates of recirculation and
afterburner costs is consequently poor. Recirculation costs, as seen in
Table III-11, are about .04 m/kwhr. Assuming that afterburners effectively
raise the temperature of the flue gas by about 40°, this would cost about
.3 m/kwhr. Both of these methods might reduce carbon monoxide emissions
by about one-half.

Table III-11

Cost and Collection Ability of Nitric Oxide Control Technologies[a]

Technology	Collection Ability[b] (%)	Capital Cost ($/kilowatt)	Operating Cost (mills/kwhr)	Annualized Cost (mills/kwhr)
Staged Combustion	35	0	0	0 (.5)
Flue Gas Recirculation	10	3	.018	.036 (1.0)
$Mg(OH)_2$ Scrubbing	87	12	.7	1.1 (2)
H_2SO_4 Scrubbing (92% Scrubber Effluent)	93	17	1.7	2.2 (3)
Limewater Scrubbing	88	9	1.5	1.8 (3)
NH_3 Reduction and Scrubbing	84	15	2.4	2.9 (3)
H_2S Reduction/Claus Process	86	12	2.4	2.8 (3)
Catalytic Decomposition	86	4	2.4	2.5 (3)

[a] Source: Table III-5 and Bartok et al. [1].

Assumptions: Figures presented are the averages of the two sources.

[b] Collective ability measured in percent of mass removed.

Alternative Control Methods

In addition to abatement techniques which are designed to reduce
the quantity of emissions, there are several other approaches which may
effectively lower air pollution damages. For instance, higher thermal
efficiency in power production reduces the environmental damage associated
with a unit of electricity. Modern plants can achieve thermal efficiencies
of 40% compared to the 1910 average of 10%. More efficient combustion
technologies and turbines permit less fuel to be burned to make each unit
of electricity. Emissions per kilowatt hour are consequently lowered.
In this sense, new turbine designs are a potential pollution abatement
method.[10] A turbine which increases thermal efficiency from 40 to 45%
would reduce across the board emissions per kwhr by about 11%. Thermal
efficiency is also important when discussing old power plants. If an
aging power plant has a thermal efficiency of 20%, it emits roughly
twice as much pollution per kwhr as a modern plant solely because of the
lower thermal efficiency. Old boilers probably also emit more pollution
per unit of fuel burned. The prolonged use of old inefficient boilers
has important environmental repercussions.

Another method of reducing the damages of dangerous emissions is to
take advantage of atmospheric dispersion. Damages can be reduced with
dispersion techniques by reducing the concentrations of pollutants in
dense populated areas at the expense of more remote areas. The benefits
depend upon whether the rural damage is greater or less than the urban
damage. Presumably, this differential will increase with the relative
density of the urban area. Siting power plants downwind of population
centers, placing them far away from cities, building higher stacks, and

reheating the flue gases are all methods of dispersing pollutants and reducing their expected damages.

For a plant located in the central city, increasing the effective height of emissions results in substantial reductions in population exposures to emissions. Table III-12 shows how population exposures may vary with effective emission heights. For a central city power plant, increasing effective heights from 20 meters to 200 meters can lower population exposures to a primary pollutant to slightly more than 1% and exposures to a secondary pollutant to about 70%. This technique primarily reduces the exposures of people in the immediate vicinity of the plant. The benefits of using higher stacks are consequently greater for plants located in central cities than in remote locations.

One method of increasing effective stack height is simply to build higher stacks. The construction cost of a higher stack is roughly $1000/ft. A taller stack may result in increased maintenance and operating costs but these costs are judged to be negligible.[11] High stacks may also be the source of new externalities. Some people may object aesthetically to higher stacks. Owners of private planes or local airports may also object to tall stacks because they interfere with low flying objects. Except for tall stacks near airport runways, these problems should be minimal.

Another method of increasing effective stack height is to heat the flue gases as they leave the plant. If the flue gases are hotter than the air around them, they will tend to rise. The following Holland [14] equation describes this increase in the effective height of the plume. It should be noted that the plume must travel several hundred meters before it will reach the height described in the equation (ΔH).

$$\Delta H = \frac{V_s d}{u} (1.5 + 2.68 \times 10^{-3} \, p(\frac{T_s - T_a}{T_s}))$$

V_s = stack gas exit velocity (m/sec)
d = inside diameter of the stack (m)
u = wind speed (m/sec)
p = atmospheric pressure (mb)
T_s = stack gas temperature (K°)
T_a = air temperature (K°)

Assuming that atmospheric pressure is about 920 mb, air temperature is about 60°F, winds are about 3 m/sec, stack diameter is 14 ft., and the stack gas velocity is about 45 ft/sec, one can analyze the relation between stack gas temperature and effective stack height. Table III-13 demonstrates the effect of higher flue gas temperatures on the effective height of the stack given the assumptions just described. For a typical power plant, it takes about 1% of the total fuel used to generate electricity in order to increase stack gas temperature by 22°K. For a 500 MW plant operating 6500 hours a year, this would represent about $25,000/yr per K°. Table III-13 also illustrates the cost of achieving higher emission heights by heating the flue gas. The marginal cost curve for this approach is clearly rising.

Another major dispersion method is choosing appropriate sites for new power plants. While there are many variables which limit siting such as access to fuel and cooling water, at least in Connecticut, many sites fulfill these conditions. The major cost among power plant locations is the cost of transmitting the power back to the load center. If a utility is not concerned with dispersing their pollutants, the cost minimizing strategy is to locate the plant as close to the load center as

63.

Table III-12

Person Exposures from SO_2 Emissions at
Several Effective Emission Heights[a]

Effective Heights (meters)	Total Person Exposures to: SO_2 SO_4 (person-$\mu g/m^3$)		Percent of Exposures in New Haven SO_2 SO_4	
20	328627	5821	98.7	22.0
40	187868	5055	97.8	10.5
60	103252	4735	96.1	5.0
80	55242	4591	92.9	2.8
100	28961	4490	87.2	1.5
120	16048	4412	78.2	.9
140	9806	4342	66.4	.5
160	6561	4277	52.9	.3
180	4742	4213	38.9	.2
200	3682	4149	26.0	.1

[a] The plant is located in New Haven harbor. The benefits
of higher stacks are considerably lower for more
decentralized plants. Emissions are equal to 1000 tons
per year. Transformation rate from sulfur dioxide (SO2)
to sulfate (SO4) is assumed to be 5%/hour.

Table III-13

The Effect of Flue Gas Temperatures on Effective Height[a]

Flue Gas Temperature (K°)	Effective Height Above Stack (meters)	Total Cost ($10³/year)	Marginal Cost ($/meter/year)
288	0	0	–
300	7.7	300	39,000
325	22.9	925	41,000
350	35.9	1550	48,000
375	47.1	2175	56,000
400	57.0	2800	68,000
450	73.4	4050	75,000
500	86.5	5300	96,000
600	106.3	7800	128,000

[a] These estimates are calculated from a Holland [14] equation with the expected parameters discussed in the text.

possible. This usually entails placing the plant in the center city.
If a utility is concerned about dispersion, it must choose both the best
direction and the best distance from the load center. The best direction
is presumably downwind from population centers. However, because
Connecticut is heavily populated, choosing the best direction is more
complicated. As one moves away from New Haven in different directions,
one is moving closer to another population center. The optimal direction
for a plant located near New Haven may be very different from the best
direction for a plant located 50 kilometers away. Locating a plant east
is generally the best direction near New Haven. This is because of the
absence of winds from the east and the relatively heavy concentrations of
people near the western end of the state. The choice of direction is
assumed not to entail any additional costs.

There are at least two costs in locating a plant far from the load
center: transmission installation costs and electricity losses in the
lines. Spending slightly more for installation of high voltage lines can
significantly reduce the percent of electricity lost in transmission. It
is assumed in this analysis that this choice is maximized and the
appropriate high voltage lines are installed (usually 345KV lines).
Several types of transmission lines are reviewed in the analysis. The
cheapest method is to use overhead lines. The annualized installation
costs of overhead lines is estimated to be $55,000 per kilometer.
Overhead lines are considered to be a blemish on landscapes by some
citizens, however, and so the real cost to society of overhead lines may
be higher than these out of pocket costs. Underground cable is certainly
more aesthetically pleasing but it costs at least $176,000 per kilometer
to install. Some of the more technically advanced underground cable have

even higher annual costs (see USDOE[26]). For instance, 300KV
superconducting DC underground cable, which is refrigerated for lower
electrical losses, costs about $300,000 a year per kilometer just to
install. In addition to the installation costs, one also faces
transmission losses. A 345KV overhead line loses about .035% of its
current per kilometer. The more advanced cables lose at most .025% of
their current per kilometer. For a 500 MW plant, operating at 6500 hours/
year, and with electricity valued at 2.2¢/kwhr, these electrical losses
translate into $25,000/km/year for an overhead line and $17,900/km/year
for a more sophisticated cable. The annualized costs of moving
electricity from a 500 MW plant are about $80,000/km/year for overhead
lines, $201,000/km/year for normal underground cable, and $318,000/km/
year for sophisticated underground cable.

Another factor in transmission line costs involves reliability.
Even though transmission lines rarely fail, one generally would not want
to trust a single line to carry all the electricity from a power plant.
One method to reduce the risks is simply to construct two independent
lines which doubles the costs. In Connecticut and in most of the
urbanized United States, however, there already exists a myriad grid of
high voltage lines which interconnect all the electrical systems. Since
any plant in Connecticut would be relatively close to this grid, a high
voltage line to this grid would probably serve as an adequate backup to
the main transmission lines back to the load center. The question of
reliability of the transmission lines is assumed to be satisfied by the
currently existing grid of high voltage lines.

Table III-14 displays the impact upon population exposures of
choosing different sites to emit pollutants. Moving a plant east of

Table III-14

The Effect of the Emission Site
upon Total Person Exposures to Sulfur Pollutants[a]

Distance East of New Haven (kilometers)	Total Person Exposures to:		Annualized Transmission Costs[b] (mills/kwhr)	Percent of Exposures to New Haven Residents:	
	SO_2	SO_4 (person-$\mu g/m^3$)		SO_2	SO_4
0	28961	4490	0	87.2	1.5
5	8558	4537	.12	43.4	4.4
10	3302	4286	.25	22.3	2.9
15	2080	4142	.37	23.7	2.6
25	1435	3679	.62	11.5	1.9
50	892	3000	1.24	4.1	1.2
75	541	2584	1.86	2.4	.8
100	576[c]	2516	2.48	1.3	.6

[a]Assumptions: 500 MW plant, 6500 hrs/yr, 100 meter stack height.

[b]Overhead transmission lines.

[c]The increase at this distance is due to proximity to other population
centers east of New Haven.

New Haven can lower the population exposures from primary pollutants
dramatically. Secondary pollutant exposures will also drop in response
to distance but not as rapidly. Dispersion works in this case because
high pollution concentrations in remote areas effect fewer people. The
major beneficiaries of remote siting are the people in the densely
populated New Haven area.

Footnotes

1 TVA has recently announced plans to build a full sized demonstration plant near its Paradise steam generator.

2 For a technical description of chemical coal cleaning process, see Hall et al. [10].

3 See National Academy of Sciences [18], p. 378.

4 See Castaldini [3] and National Academy of Sciences [18], Appendix 11A, for a description of the possible effects of boiler operations upon sulfate and sulfur trioxide emissions.

5 The estimates in Table 5 assume sludge disposal is handled onsite with clay lined ponds. Reliable closed loop operation has not yet been demonstrated so that liquid disposal is a serious problem with these throwaway systems.

6 Adjustments for different year estimates have been corrected with the CPI.

7 Bechtel [2] mentions variations in SO_3 emissions from FGD systems. Additional evidence on particulates suggests that if the demister is working improperly, the scrubber may emit large quantities of small particles which could easily include sulfates.

8 See the discussion on health effects in Chapter V for a review of the evidence on the toxicity of each element.

9 See Noll and Davis [19] for a review of electrostatic precipitator designs.

10 See the National Academy of Sciences [18] for a discussion of new turbines.

11 This information was gathered from discussions with Jim Crowe, chief mechanical engineer at United Illuminated.

Bibliography

1. Bartok, W.; Crawford, A.R.; and Skopp, A. "Control of NO_x Emissions From Stationary Sources." Chemical Engineering Progress, 67 (February, 1971), 64-72.

2. Bechtel Corporation. Flue Gas Desulfurization Systems: Design and Operating Parameters, SO_2 Removal Capabilities, Coal Properties, and Reheat. San Francisco: November, 1977.

3. Castaldini, et al. Boiler Design and Operating Variables Affecting Uncontrolled Sulfur Emissions from Pulverized Coal-Fired Steam Generators. Mt. View, California: Acurex Corporation, October, 1977.

4. "Combustion, Pollution Controls." Power (April, 1974), 49-64.

5. Devitt; Yecino; Ponder; and Chatlyne. "Estimating Costs of FGD Systems for Utility Boilers." JAPCA, 26 (March, 1976), 204-209.

6. Drehmel, D.C. "Fine Particle Control Technology: Conventional and Novel Devices." JAPCA, 27 (February, 1977), 138-140.

7   Engdahl, R.B. "The Status of Flue Gas Desulfurization." Air Pollution Central Division News, 5 (April, 1977).

8. Federal Power Commission. Staff Report of Bureau of Power. The State of Flue Gas Desulfurization Applications in the United States: A Technological Assessment. July, 1977.

9. GCA Corporation. Status Report on Control of Particulate Emissions From Coal-Fired Utility Boilers. Bedford, Mass.: 1977.

10. Hall; Hoffman; Hoffman; and Schilling. Evaluation of Physical Coal Cleaning as an SO_2 Emission Control Technique. Columbus, Ohio: Battelle Memorial Institute, 1977.

11. Haller, G.L. and Nordine, P.C. "Combined Coal Cleaning with Wet Lime (Limestone) Flue Gas Desulfurization." Department of Engineering and Applied Science, Yale University, May, 1975.

12. Herling, J. Flue Gas Desulfurization in Power Plants: A Status Report. U.S.E.P.A., Washington, D.C.: April, 1977.

13. Hoffman, L. and Deurbruck, A.W. Engineering/Economic Analyses of Coal Preparation With SO_2 Clean-Up Processes For Keeping Higher Sulfur Coals In The Energy Market. Proceedings of Physical Coal Cleaning Conference, Louisville, Ky., 1976.

14. Holland, J.Z. A Meteorological Survey of the Oak Ridge Area.
    Atomic Energy Commission Report, ORO-99, Washington, D.C.,
    1952.

15. Industrial Gas Cleaning Institute. A Briefing on SO_2 Control.
    Stamford, Conn.

16. McGlamery, G.G. and Tolstrick, R.L. Cost Comparison of Flue Gas
    Desulfurization Systems. TVA, Knoxville, Tenn., 1974.

17. National Academy of Engineering. Abatement of Sulfur Oxide Emissions
    from Stationary Combustion Sources. Washington, D.C., 1970.

18. National Academy of Sciences. Air Quality and Stationary Source
    Control. Report to Committee on Public Works, U.S. Senate,
    Washington, D.C., 1975.

19. Noll, K.E. and Davis, W.T. Power Generation Air Pollution Monitoring
    and Control. Ann Arbor Science, Ann Arbor, Mi., 1976.

20. Offen, G.R.; Kesselring, J.P.; Lee, K.; Poe, G.; and Wolfe, K.J.
    Control of Particulate Matter from Oil Burners and Boilers.
    EPA-450/3-76-005, April, 1976.

21. Ondov, J.M.; Ragaini, R.C.; and Biermann, A.H. "Wet Scrubber vs.
    Electrostatic Precipitators: Relative Particulate Inhalation
    Hazards for Toxic Species. Lawrence Livermore Laboratory,
    Livermore, Calif., July, 1976.

22. Patskov, E.A.; Rozenfeld, E.I.; and Federov, V.A. "Reducing the
    Formation of NOX in Power Station Boilers and in Industrial
    Gas-Fired Furnaces." Teploenergetika, 21 (1974), 55-60.

23. Rosoff, J. and Rossi, R.C. Disposal of By-Products from Non-
    Regenerable Flue Gas Desulfurization Systems. EPA-650,
    Aerospace Corporation, May, 1974.

24. Shannon, L.J.; Gorman, P.G.; and Park, W. Feasibility of Emission
    Standards Based on Particle Size. EPA-600/5-74-007, March,
    1974.

25. U.S. Department of Commerce. Report on Sulfur Oxide Control
    Technology. Commerce Technical Advisory Board, 1975.

26. U.S. Department of Energy. Evaluation of the Economical and
    Technological Viability of Various Underground Transmission
    Systems for Long Feed Areas to Urban Load Areas. HCP/T-2055/1,
    December, 1977.

27. U.S. Environmental Protection Agency. Draft of Standards: Support and Environmental Impact Statement Proposed Standards of Performance for Electric Utility Steam Generating Units (Nitrogen Oxides and Particulates). Washington, D.C., December, 1977.

CHAPTER IV
ATMOSPHERIC TRANSPORT AND DISPERSION

This chapter discusses the movement and alterations of pollutants from the power plant stack to ground levels in surrounding locations. Meteorological factors such as the wind direction, wind speed, cloud cover, sunshine, and temperature gradients are all taken into account with a Gaussian plume dispersion model. Chemical alterations, dry deposition, and wet deposition are estimated in various subroutines which have been appended to the basic dispersion model. These routines estimate the final ground level concentrations of several pollutants within a wide area around the emission site.

The focus of this modelling is upon exploring the dispersion of pollutants within relatively short ranges (less than 100 kilometers). The Gaussian plume model used in this study is reasonably accurate for such distances. Even the best atmospheric dispersion models, however, make sizeable errors. The ground level concentration could be off by as much as a factor of three. The effects of small changes in location of emitters (between 5 and 50 kilometers), nonetheless, can be captured by this meteorological model. Because some pollutants travel considerably further than 100 kilometers, gross estimates of exposures up to 500 kilometers from New Haven are also calculated (beyond this distance, one is primarily over the Atlantic Ocean). These long range calculations, unfortunately, are quite crude. The lack of accuracy of the long range estimates in this study, however, should not seriously taint the validity of the results because long range transport does not pose a serious problem for emissions from New England and because none of the abatement strategies considered here attempt to take advantage of long range transport.

Another important focus of the model is upon annual or long term as opposed to hourly or acute dosages of pollutants.[1] Since all of the abatement methods considered here are independent of daily or short term weather conditions, it is not necessary to have a model which predicts short term concentrations of pollutants.[2] The Gaussian plume model is used here because it is relatively simple and because it predicts annual concentrations of pollutants with reasonable effectiveness.

The Gaussian plume, which simulates the dispersion of gases and airborne materials, has been developed and calibrated by Pasquill [9], Briggs [2], Gifford [4], and others.[3] The atmospheric dispersion model assumes that airborne substances distribute themselves normally around a centerline which is determined by the direction of the wind. The vertical distribution is a function of the distance from the source and atmospheric stability conditions (which are a function of the temperature gradient). Horizontal dispersion occurs within each of 16 equal sectors and is assumed to be uniform within each sector.[4] The prevalence of certain wind directions is captured by the observed frequency of winds from that direction. Ambient concentrations $\chi(x,\theta)$ at a given distance (x) and direction or sector (θ) are distributed according to the following formula:

$$\chi(x,\theta) = \frac{2\ Q\ f(\theta,s)}{\sqrt{2\pi}\ \sigma_z(s)\ u\ \frac{(2\pi x)}{16}}\ e^{-\frac{1}{2}\left(\frac{H}{\sigma_z(s)}\right)^2}$$

Q is the continuous rate of emission, s is the stability-wind speed condition, u is the wind speed, H is the effective height of release of the effluent, $\sigma_z(s)$ is the standard deviation of vertical dispersion, and

$f(\theta,s)$ is the probability of stability condition s occurring with a wind from the θ direction. An important advantage of this model is that it can be calibrated with data commonly recorded by major airports. The necessary data include a wind rose table which describes the frequency of wind speeds by sixteen directions and a table of the frequency of sky cover. This data is used to generate a frequency of twenty-four potential weather conditions, four wind speeds by six stability classes.

The most complicated aspect of generating the twenty-four weather conditions is determining stability classes. Six stability classes are used in this analysis to describe the ability of the effluent to disperse vertically. The more stable the atmosphere, the more restrictive the vertical dispersion becomes. The stability class, or Pasquill type, is a function of the vertical temperature gradient and the wind speed which can be estimated with measures of sunlight, cloud type and amount. The following rules are extracted from Turner [10] as objective measures of Pasquill types. The rules have been slightly altered to match the available data.

1. If the total cloud cover is 8-10 and the ceiling is less than 5000 feet, use a net radiation index equal to zero (day or night).

2. For nighttime:
   a. If total cloud cover is less than 4/10, use a net radiation index equal to -2.
   b. If total cloud cover is greater than 4/10, use a net radiation index equal to -1.

3. For daytime:
   a. Add the insolation class number as a function of solar altitude (a): $a > 60$, then 4; $60 > a > 35$, then 3; $35 > a > 15$, then 2; $15 > a$, then 1.
   b. If total cloud cover is < 4/10, add zero.
   c. If total cloud cover is $8/10 > C \geq 4/10$, and the ceiling is < 5000, add -2.

d.   If total cloud cover is 8/10 > $C \geq$ 4/10, and the ceiling
     is > 5000, add --1.
e.   If total cloud cover is $\geq$ 8/10, and the ceiling is $\geq$ 5000,
     add -1.
f.   If the modified insolation class is $\leq$ 1, it is 1.
g.   The stability index is a function of the modified
     insolation (net radiation index) and the wind speed
     (see Table 2).

The stability numbers of Table IV-1 directly correspond to the

letter categories established by Pasquill [9]. Using formulas developed

by Briggs [2] for urban settings, the vertical dispersion coefficients

are computed from the stability classification and the distance from the

original source (x). These formulas are shown in Table IV-2. The most

unstable classification, A (1) corresponds to the greatest vertical

dispersion and consequently the lowest ground level concentrations.

Table IV-1

Stability Classifications

Wind Speed	Net Radiation Index						
	4	3	2	1	0	-1	-2
0-3	1	2	2	3	4	5	6
4-7	2	2	3	4	4	4	5
8-12	3	3	4	4	4	4	4
13+	3	4	4	4	4	4	4

Table IV-2

Formulas for the Vertical Dispersion Coefficients[a]

Stability Class	Pasquill Type	Formula
1	A	$.24 \, x \, (1 + .001 \, x)^{\frac{1}{2}}$
2	B	$.24 \, x \, (1 + .001 \, x)^{\frac{1}{2}}$
3	C	$.20 \, x$
4	D	$.14 \, x \, (1 + .003 \, x)^{-\frac{1}{2}}$
5	E	$.08 \, x \, (1 + .0015 \, x)^{-\frac{1}{2}}$
6	F	$.08 \, x \, (1 + .0015 \, x)^{-\frac{1}{2}}$

[a]
See Briggs [2].

A frequency table of the twenty-four possible weather conditions $f(\theta,s)$ is constructed in the following manner. Six stability types are cross classified with four wind speed categories. Since the data used in this analysis did not provide joint distributions of wind direction, cloud cover, ceiling height, and wind speed, both cloud cover and ceiling height are assumed to be independently distributed across all wind directions. Though either of these factors may be correlated with wind direction, the assumption of independence should result in only small errors.

For large distances, one must consider the possible interference of ceiling heights upon vertical dispersion. The model assumes that, for the first eighty kilometers, the pollutants can disperse according to the Gaussian plume model. Beyond eighty kilometers, no additional vertical dispersion is modelled; the only dispersion which occurs is in the horizontal direction (as the distance increases, each sector gets wider). Restricting vertical dispersion beyond eighty kilometers is somewhat arbitrary. It appears that the plume will strike the mixing height well before eighty kilometers. Vertical dispersion after this period is clearly restricted but it is not necessarily completely prevented.

In the crude Gaussian plume model, particles never settle to the ground and the plume disperses forever into an infinite plane. Wet and dry deposition, however, are important phenomena to include in an atmospheric transmission model. The fraction of material removed by wet deposition is assumed to be proportional to the amount of time the particle remains in the atmosphere and the average rate of precipitation. The average frequency of rainfall overestimates true wet deposition because rain often falls through an atmosphere which has already been cleansed by previous hours of rain. Wendell [12] estimates that using average precipitation rates overestimates real wet deposition by about 20%. In order to correct this error, the fraction of material removed by wet deposition is multiplied by .80. The fraction of material removed by wet deposition by a certain distance x is:

$$f_w(x) = W_i * (x/u) * .80$$

where u is the wind speed and W_i reflects the rate of removal of pollutant i. The coefficient W_i depends on the precipitation rate, the drop size, and the collection efficiencies for each pollutant in the following manner:

W_i = Precipitation rate * collection efficiency/drop rate

Utilizing the average precipitation rate for the coastal region around New Haven (130 mm/year) and assuming an average drop size of 2mm, one can calculate the wet deposition rates for a number of air pollutants (see Table IV-3).

Table IV-3

Wet Deposition Rates for Southern Connecticut (W_i) [a]

Pollutants	Expected	Standard Deviation
SO2, NO2, TSP	.005	.001
SO2, NO2	.0005	.0002

[a]Differences among pollutants are due to different collection efficiencies.

Dry deposition, the settling out of particles because of gravity and of gases because of adsorption, depends on the time a particle remains in the atmosphere, the particle's average deposition velocity, and the effective height of the plume. The effective height of the plume is related to the plume concentration (χ/Q). The fraction of the plume removed by dry deposition (f_d) depends upon the deposition velocity (D_i), the length of the discrete distance under consideration (Δx), the total distance from the source (x), and the density of the plume:

$$f_d = D_i * \Delta x * \chi(x) * \left(\frac{8x}{Q}\right)$$

The deposition velocity of the particle (D_i) changes with different ground cover. According to estimates by Gudicksen [5], the following ground covers adsorb pollutants by increasing amounts: agricultural

land, grassland, brushland, and forest. The ambient concentration from
an equivalent source after passing over a forest may be less than one-
third of what it would be after passing over agricultural land. Polluters
located in large forested tracts will have less of an impact on distant
locations than polluters located near agricultural land.

Chemical reactions which occur while the emittant is in transit
significantly alter the mix of pollutants over even intermediate
distances. Radioactive materials tend to decompose into less harmful
daughter products. Fossil fuel effluents, on the other hand, tend to
oxidize into more harmful acidic compounds. Nitrogen oxide changes into
nitrate or oxidants and sulfur dioxide becomes sulfate. The model
assumes that these changes are merely a function of time and that they
occur at a constant rate of change. More complex versions of these
processes do not appear to be warranted given current uncertainty about
the processes and the relative difficulty of incorporating more detailed
interactions. The results of some complex interactions can, nevertheless,
be simulated with the simple model used here. For instance, the possible
qualitative effect of additional levels of hydrocarbons as a catalyst in
the conversion of SO_2 to SO_4 can be explored by trying higher rates of
conversion. The impact of higher levels of background hydrocarbons can
be evaluated in terms of increases in the population exposure to more
rapidly formed sulfate. Efforts to include these interations, however,
should be considered exploratory since little is known about the
magnitudes of the effects.

The rate of transformation of SO_2 to SO_4 is a very complex and
poorly understood parameter. Heat, sunlight, relative humidity, ozone,
hydrocarbons, and nitrogen oxides are all suspected of altering the rate

of transformation. Values in the literature range from .5 to 50% per hour.[5] The relatively higher figures reported for urban areas are used in this study. The expected rate of change of sulfur dioxide to sulfate is assumed to be 5% per hour and the standard deviation of this parameter is about 2%.

The transformation of NO to NO_2 occurs at the relatively rapid rate of about 15% per hour (see University of Washington [11]). Less is known, however, about the rate of transformation of NO_2 to nitrate or oxidants. Light appears to play an important role in the formation of oxidants. Ozone formation is also likely to be more rapid with high humidity or concentrations of hydrocarbons. The evidence of these secondary reactions in the wake of power plant plumes, though, contradicts itself. Davis [3] reports an increase in ozone in a power plant plume. University of Washington [11] and Meteorology Research Inc. [7] find no evidence of ozone formation in plumes in isolated western power plants. Because ozone formation is more likely to occur in more polluted environments, plumes in isolated areas are less likely to show any effects. The western evidence only provides a lower bound on the effect of an eastern power plant plume on ozone formation. It is assumed in this paper that oxidants are formed out of the nitrogen oxides from a power plant at a constant and slow rate of 1% per hour. The uncertainty around this estimate is quite large and one should be cautious applying these model results. Nitrate is assumed to be formed at the same rate of speed as the transformation from NO to NO_2 (see NAS [8]).

Because wet and dry deposition and chemical change all occur simultaneously in this discrete model, the final fraction of material removed (f_T) depends on the interaction of each removal process:

$$f_T = f_k + (1 - f_k) f_w + (1 - f_k) (1 - f_w) f_c$$

Rearranging these terms leaves:

$$f_T = f_d + f_w + f_c - f_d f_w - f_d f_w - f_w f_c - f_d f_w f_c$$

Summing over all periods, the total fraction of material $(F_T(x))$ remaining by distance x is:

$$F_T (x) = e^{-\Sigma f_T}$$

$$F_T (x) = e^{(-\Sigma f_d - \Sigma f_c - \Sigma f_w + \Sigma f_d f_w + \Sigma f_d f_c - \Sigma f_d f_c f_w)}$$

The final quantity of material which is still in the air at distance x from the original source is the product of the amount emitted times the fraction remaining $F_T(x)$. The final ambient concentration of material by distance x $(Amb(x))$ is the product of the Gaussian dispersion $(\chi(x))$ and the fraction remaining.

$$Amb (x) = \chi(x) * F_T (x)$$

It is quite difficult to accurately account for the ambient concentrations of the sulfate and nitrate which is formed in the atmosphere. As soon as these compounds oxidize, they acquire lower deposition rates. Each of the sulfate particles formed from sulfur dioxide (and similarly for nitrates from nitrogen dioxide) enter the atmosphere as sulfates at different times and locations (along the path of the plume). These new particles are distributed differently from the original sulfur dioxide and from any original sulfate emissions. As a rough approximation, it is assumed that all these particles are distributed as if they had been emitted as sulfates (or nitrates) except that they are subject to deposition rates for only one half the distance from the origin to the final location (since, on average, they are created half way between these points). In order to calculate the ambient concentrations of sulfates and nitrates which are formed in the atmosphere, one merely has to calculate the total quantity of sulfur dioxide or nitrogen dioxide which has been converted by chemical change and treat that quantity as if it were an original emission of sulfate or nitrate except for the lower deposition rate.

In order to understand more about the relationship between the parameters of the model and the model's predictions, several values of each parameter are studied in a sensitivity analysis. The parameters for wet deposition, chemical change, and dry deposition are examined separately in Tables IV-4, IV-5, and IV-6, respectively. Though it is difficult to be certain about the range within which the true parameters must lie, the parameters shown in Tables IV-4 through IV-6 are intended to span a plausible range of values. The primary output of the dispersion model is the total population exposure which results from an emission. The emission measured in these examples originates in central New Haven. The distribution of these exposures by distance from the source is also presented.

Tables IV-4 and IV-6 demonstrate that additional wet or dry deposition may lower total exposures to the population. Greater dispersion rates especially reduce distant exposures by dampening long distance transport. Thus, as the deposition rate increases, a greater percentage of the remaining exposures are located near the emission source. However, changes in the wet deposition rate (about the size of observed annual

Table IV-4

The Effect of Wet Deposition Rates
on Person Exposures from an Emission of SO_2[a]

Rate of Wet Deposition (%/hr)

	SO_2	SO_4	SO_2	SO_4	SO_2	SO_4
	.002	.0002	.007	.0007	.021	.0021
Total Persons Exposed	29070	4881	28961	4490	28697	3664

Percent Distribution by Location

New Haven	86.9	1.4	87.2	1.5	87.9	1.9
< 10 km	8.6	4.3	8.5	4.6	8.4	5.5
11-25	.8	2.2	.8	2.3	.7	2.6
26-50	.9	6.9	.9	7.1	.8	7.6
51-100	1.9	39.9	1.8	40.0	1.5	40.1
101-200	.7	31.1	.7	30.7	.5	29.7
201-300	.1	9.7	.1	9.4	.1	8.8
301-500	.0	4.4	.0	4.2	.0	3.7

[a]Emission in New Haven, effective height of 100 meters, 1000 tons/year and all other parameters set to expected values.

Table IV-5

The Effect of the Rate of Chemical Change
on Person Exposures from an Emission of SO_2[a]

	Rate of Transformation (%/hr)					
	SO_2	SO_4	SO_2	SO_4	SO_2	SO_4
	1		5		25	
Total Persons Exposed	30784	1827	28961	4490	27105	6936
Percent Distribution by Location						
New Haven	82.2	.7	87.2	1.5	92.1	4.9
< 10 km	8.5	2.4	8.5	4.6	6.9	11.6
11–25	1.0	1.4	.8	2.3	.3	3.8
26–50	1.5	4.9	.9	7.1	.2	8.2
51–100	4.3	37.9	1.8	40.0	.3	35.8
101–200	2.0	34.6	.7	30.7	.1	25.2
201–300	.4	12.1	.1	9.4	.0	7.3
301–500	.1	5.9	.0	4.2	.0	3.2

[a]Emission in New Haven, effective height of 100 meters, 1000 tons/year and all other parameters set to expected value.

Table IV-6

The Effect of Dry Deposition Rates
on Person Exposures from an Emission of SO_2[a]

	Dry Deposition Rate ($\%$/hr)					
	SO_2	SO_4	SO_2	SO_4	SO_2	SO_4
	.01	.001	.05	.005	.25	.025
Total Persons Exposed	29035	4665	28961	4490	28630	3347

Percent Distribution by Distance from New Haven

	SO_2	SO_4	SO_2	SO_4	SO_2	SO_4
New Haven	86.9	1.5	87.2	1.5	88.1	2.0
< 10 km	8.6	4.5	8.5	4.6	8.2	5.8
11-25	.8	2.3	.8	2.3	.7	2.6
26-50	.9	7.1	.9	7.1	.7	7.7
51-100	1.9	40.2	1.8	40.0	1.6	40.3
101-200	.7	30.7	.7	30.7	.6	29.5
201-300	.1	9.4	.1	9.4	.1	8.5
301-500	.0	4.2	0.0	4.2	.0	3.5

[a]Emissions in New Haven, effective height of 100 meters, 1000 tons/year and all other parameters set to expected value.

changes in rainfall) result in only minute changes of total population exposures. A 50% increase in rainfall would increase the wet deposition rate by about .008%/hr and would decrease total population exposure from an emission of SO_2 in New Haven by less than 10%. Changes in dry deposition rates also have small effects on total population exposures. Doubling the rate of dry deposition decreases total exposures by less than 5%.

Population exposures are particularly sensitive to the rate of chemical change of sulfur dioxide to sulfate. The total sulfate exposure from a unit of sulfur dioxide emission doubles with a 5% rather than 1% rate of transformation per hour. The sensitivity of the chemical change parameter is clearly important. Given the current uncertainty about its true value, the variance of the parameter will substantially increase the variance of our final understanding of the dispersion of sulfur dioxide. An important research priority in atmospheric modelling is to ascertain a better understanding of these chemical changes. Also, controlling the levels of certain background pollutants which may increase this rate of transformation may be justifiable solely because of their deleterious impact on ambient sulfur products.

Footnotes

[1] The emphasis in this study upon long term effects is based on epidemiological evidence which suggests that chronic exposures are far more damaging than acute exposures and can account for over 90% of health losses. This evidence is reviewed in Chapter V.

[2] Lange [6] describes the pitfalls of using Gaussian plume models for short term predictions.

[3] For a recent review of the literature, see Gifford [4]. Another good source for practical meteorological models is Environmental Protection Agency, Workbook of Atmospheric Dispersion Estimates, Research Triangle Park, North Carolina, 1970.

[4] The 16 equal sector model was developed by J.Z. Holland, "A Meteorological Survey of the Oak Ridge Area", Atomic Energy Commission Report, ORO-99, Washington, D.C.

[5] See the discussion on transformation rates in Alkezweeney [1] and NAS [8], Chapter 6.

Bibliography

1. Alkezweeny, A.J. and Powell, D.C. "Estimates of The Transformation Rate of Sulfur Dioxide to Sulfate From Atmospheric Concentration Data." Atmospheric Environment, XI (1977), 179-82.

2. Briggs, G.A. Diffusion Estimation for Small Emissions. In Air Resources Atmospheric Turbulence and Diffusion Laboratory 1973 Annual Report. USAEC Report ATDL-106, December, 1974.

3. Davis, D.D.; Smith G.; and Klauber, G. "Trace Gas Analysis of Power Plant Plumes Via Aircraft Measurement: O_3, NO_x, and SO_2." Science, 186 (1974), 733-36.

4. Gifford, F.A. "Turbulent Diffusion-Typing Schemes: A Review." Nuclear Safety, XVII (1976), 68-86.

5. Gudicksen, P.H.; Peterson, K.R.; Lange, R.; and Knox, J.B. Plume Depletion Following Postulated Plutonium Dioxide Releases from Mixed Oxide Fuel Fabrication Plants. Lawrence Livermore Laboratory, UCRL-51781, 1975.

6. Lange, R.; Dickerson, M.A.; Peterson, K.R.; Sherman, C.A.; and Sullivan, T.J. Particle in Cell vs. Straight Line Airflow Gaussian Calculations of Concentration and Deposition of Airborne Emissions. Lawrence Livermore Laboratory, UCRL-52133.

7. Meteorology Research, Inc. Oxidant Measurements in Western Power Plant Plumes. EPRI EA-421, 1977.

8. National Academy of Sciences. Air Quality and Stationary Source Emission Control. Report prepared for the Senate Committee on Public Works. Washington, D.C.: U.S. Government Printing Office, 1975.

9. Pasquill, F. "The estimation of the dispersion of windblown material." Meteorological Magazine, 90 (1961), 33-49.

10. Turner, D.B. "Relationships Between 24 Hour Mean Air Quality Measurements and Meteorological Factors in Nashville, Tennessee." Journal of Air Pollution Control Association, (1961).

11. University of Washington. Reactions of Nitrogen Oxides, Ozone, and Sulfur in Power Plant Plumes. EPRI EA-270, 1976.

12. Wendell, L.L.; Powell, D.C.; and Drake, R.L. A Regional Scale Model for Computing Deposition and Ground Level Air Concentrations of SO_2 and Sulfates from Elevated and Ground Sources. Third Symposium on Atmospheric Turbulence, 1976.

CHAPTER V

DOSE RESPONSE CURVES

The weakest link in our understanding of the environmental chain between emissions and final effects lies in predicting the final effects of exposures to each air pollutant. Despite decades of research and centuries of suspicion, the type of effects, the specific agent of destruction, and the magnitude of the damages from air pollution are still poorly understood. This state of affairs can partially be blamed upon the lack of sufficient, well planned, natural and laboratory experiments, and the virtual absence of sophisticated analysis of unplanned natural events. On the other hand, because of the complexity of the environment, the costs and time constraints placed on research, and the subtle nature of some dose effects, it is encouraging that we are at least improving our understanding of dose effects over time.

It should be noted that there is not one but a whole universe of dose response curves that need to be explored. For instance, the length of time a subject is exposed to a pollutant often illicits significantly different responses. This has led to a general distinction between acute and chronic effects. The dose responses among different subjects is another important distinction. Some animal species are highly resistant to pollutants which are fatal to other species. Even within species, some animals will be effected while others will not. There also may be more than one effect to capture. For example, a specific exposure may result in excess mortality, acute respiratory disease, cough, and minor eye irritation. Further, there may be important synergistic effects among pollutants and among other environmental factors. One needs to

evaluate each pollutant under a number of plausible environmental
conditions. Dose response curves can vary depending on the mixture of
pollutants, the length of exposure, the object exposed, the environmental
conditions, and the types of effects. Existing research has only just
begun to unravel the consequences of air pollution exposures.

Ideally, in order to determine the specific consequences of air
pollution exposures upon various objects, one would want to carry out a
series of experiments in both the laboratory and the field. Trying to
hold most environmental factors at a predetermined level, one would want
to (perhaps independently) vary the length of exposure, the concentration,
the mixture of pollutants, and several other environmental factors.
Through a series of such exposures, it would be possible to isolate the
impact of particular pollutants and to explore the interactions amongst
each pollutant, different subjects, and other environmental factors.
Unfortunately, practice falls short of this ideal. As already mentioned,
there are so many relationships to explore, it would take enormous
experimental programs to understand them all. Also, some effects are so
subtle that they require large populations in order to study them. Time
and cost act as constraints upon research methods and common scientific
practice must often deviate from methods which would eventually maximize
our understanding of air pollution.

The time and cost constraints placed on research encourage the use
of large dosages, short exposure times, and relatively small samples.
These methods may be able to identify especially toxic materials, but it
is less likely they can identify moderately or chronically harmful
pollutants. Further, direct extrapolation of results from these
experiments to predictions of the effects of long term or low dose

exposures is treacherous. Animals are often more resistant to chronic
than acute doses. Acute dose response curves may overestimate chronic
effects. On the other hand, chronic effects may not even be visible in
acute experiments. With respect to the use of abnormally high dosages,
all natural and social scientists are aware of the danger of extrapolating
results beyond the region examined. With most laboratory exposures ranging
from three to four orders of magnitude above natural ambient concentrations,
quantitative estimates of relevant effects are illusive. Most laboratory
experiments only provide reliable qualitative results. For quantitative
estimates, one must rely heavily on the few planned experiments done
under normal conditions and on occasional evidence uncovered through
unplanned experiments.

A program of planned experiments clearly provides a better chance of
isolating the effects of desired treatment variables than do analyses of
uncontrolled and complex environments. Ideal experimental studies, as
just discussed, are often not possible. With planned experiments
thwarted by cost and moral constraints, unplanned experiments may provide
a valuable alternative opportunity. Expecially in the case of health
effects, where long term experiments on humans are not feasible on both
cost and moral grounds, the remaining planned experimental alternatives
offer only a remote opportunity of quantifying the human health effects
of damaging pollutants. In such cases, the opportunity to learn more
about the consequences of pollutants from naturally occurring events
should not be ignored.

The primary hazard of analyses of unplanned natural phenomena is
failure to account for variation from unwanted sources. This is
expecially disconcerting when the treatment variable is correlated with

unwanted factors because it is easy to be mistaken about the causal link between the treatment variable and the response. Even if the unwanted variations are uncorrelated with the treatment variable, these variations reduce the accuracy of the estimate of the desired parameters. Analyses of unplanned phenomena should, therefore, be designed in such a way as to minimize the impact of unwanted variables.[1] One way to accomplish this is to carefully select sample points so that serious unwanted variables are held constant across the sample. Epidemiologists have used this technique in order to control some personal characteristics in human health studies. For instance, all the subjects in a study will be of the same age and sex. Another method is to transform the data so as to control certain undesired phenomena. Analyzing time series data through the rate of change between time periods rather than absolute levels is a transformation which dampens the effect of slow changing long term factors. A third method to account for unwanted variation is to include proxies for undesirable variables. Though it is difficult to find adequate proxies and appropriate functional forms, it is at least possible to remove the most burdensome effects of unwanted factors through statistical methods.

Estimating pollution dose response curves from natural experiments, unfortunately, is further hampered by inadequate measurements of the pollution itself. Instrumental readings are often inaccurate, some pollutants are not measured, characteristics of the chemical state of the pollutants are often ignored, historical readings are rare, and personal mobility tends to erode the correlation between personal and area wide exposures.

Exploring natural phenomena has the advantage, on the other hand, of observing relationships under relevant conditions, the conditions under which people actually live. Another advantage is that the effects of long term exposures, if correctly captured, can be studied immediately since most pollutants have existed for several decades. Further, if some of the enormous data sets which are gathered for other purposes were utilized, it would be possible to observe infrequent and subtle effects which are difficult to detect in small planned experiments.

In order to evaluate the health effects of air pollution from fossil fuel power plants, it is important to borrow from the literature of both planned and unplanned experiments. The following pollutants are especially relevant to this discussion: sulfur dioxide and other sulfur products, nitrogen dioxide and other nitrogen products, oxidants, carbon monoxide, particulate matter, and hydrocarbons. The effect of each of these pollutants on aesthetics, materials, vegetation, chronic health effects, and acute health effects are discussed, in turn, throughout the remainder of this chapter.

A. Aesthetic Effects

The classification of consequences amongst aesthetic, material, and minor health damages is somewhat arbitrary. In this analysis, loss of visibility, cleaning costs, acid rain, and annoying smells are included as aesthetic losses.

Though many industrial processes result in annoying smells, the products of fossil fuel combustion rarely can be detected by the nose. Sulfur dioxide, if detected, is certainly displeasing (a small frequently associated with rotten eggs). The minimum threshold concentration for

sulfur dioxide, however, is 1600 $\mu g/m^3$. Concentrations at ground level
from power plant plumes do not approach this level. Many of the other
major products of fossil fuel plants are tasteless. Modern fossil fuel
plants, since they burn fuels relatively efficiently, are not expected
to present a major olfactory nuisance.

Several scientists have observed a correlation between loss of
visibility and ambient concentrations of particulates.[2] The particles
apparently effect visibility by refracting light. Pollution, however, is
not the only factor which reduces visibility—naturally produced dust and
water droplets (high humidity) can also refract light. Also, not all
pollutants effect visibility since some are invisible gases (carbon
oxides, nitrogen oxide, and sulfur dioxide). These invisible pollutants
contribute to visibility problems only indirectly by adhering to particles
in the atmosphere and creating harmful aerosols.

There is general agreement in the literature that particulates
diminish visibility, though at a decreasing rate with higher concentrations.
In urban environments, according to Charlson et al.[13], an additional
$\mu g/m^3$ of particulates only reduces visibility by .4 km (assuming a
background level of 70 $\mu g/m^3$). An additional unit of particulates in
rural areas may reduce visibility as much as 16 km (with a background
level of 10 $\mu g/m^3$). In Connecticut, the average background level of
particulates is 55 $\mu g/m^3$ and so an additional $\mu g/m^3$ is assumed to reduce
visibility by .6 kilometers. This marginal value is also assumed to be
the same for units of nitrogen dioxide and sulfate. The total effect of
the power plant's damage to visibility is measured in terms of the number
of miles of visibility times the number of people affected (people-
kilometers).

Another aesthetic cost to consider is excessive cleaning bills. If boilers burn inefficiently, they can produce large quantities of ash. If this ash is not removed from the flue gas, it can shower surrounding areas with considerable soot. The deposition of large quantities of such materials clearly imposes damages upon nearby residents. Ridker [40] measured the cleaning costs of an accidental release of soot from a power plant. He found that nearby residents underwent certain additional cleaning costs as a result of the incident. The residents, however, rarely valued the damages above their out of pocket costs for the cleaning. The additional psychic costs of a heavy deposition of particles appears to be negligible. British attempts, Atomic Energy Research Corporation [5], to correlate smoke with cleaning costs were generally unsuccessful. The only variable which was correlated with heavier smoke in an area was more frequent cleaning of automobiles. Given this evidence, it is unlikely that small changes in deposition rates will result in substantial losses. As a rough approximation, it is assumed that an annual deposition density of 1 gram/meter^2 would damage afflicted residents about one dollar per year. The standard error around this figure is large, at least about 6. This effect is a negligible part of total damages.

Acid rain, a result of wet deposition of sulfur and nitrogen oxides, produces subtle effects on ecological systems and materials. The severity of acid rain is thought to depend on its acidity or pH factor. The most acidic rains fall in northern New England--far away from the original emission sources. The acid rain is suspected to cause leaf damage, erode statues and monuments, slow plant growth, kill fish, and eventually damage water supplies and soils. The magnitude of these effects are not yet understood. As a rough approximation, acid rain is assumed to cause $400 million damage each year nationally.[3] Assuming that three-fourths of these damages are due to sulfur and one-fourth to nitrogen oxides, and a direct relation between emissions and total damage, the acid rain damage of every ton of sulfur emitted is about $10, and for every ton of nitrogen, it is about $4.

B. Materials

There are a number of examples where pollution clearly has damaged materials. Measuring the extent of this damage and quantifying these effects, however, can only be done cautiously given the available information. For an overview of material damages, see Waddell [52] or NAS [39].

There is strong evidence to suggest that ambient sulfur products, even at normal concentrations, corrode steel. The corrosion rate appears to vary with the level of oxidants, the temperature, and the relative humidity. Haynie and Upham [22] demonstrate in a ten year field study that the presence of oxidants dampens the rate of corrosion from sulfur products. Sereda [43] shows that higher humidity and higher temperatures (together these mean greater water vapor pressure or more water in the air) result in significantly faster corrosion rates. Because sulfate is more likely to be present at higher humidities, the damaging agent in the Sereda study may be sulfate and not sulfur dioxide. Corrosion studies with zinc indicate similar relationships. Although these studies suggest sulfur products corrode metals, the magnitude of this effect does not appear to be large. Assuming that pollution concentrations are near the

national average, an additional $\mu g/m^3$ of sulfur dioxide corrodes about
5 microns into steel and less than 1 micron into zinc in ten years. At
this rate, it would take at least 1000 years to corrode one centimeter
into either of these metals. This rate of corrosion is inconsequential
for most uses of these metals. Further, these metals are rarely exposed
directly to the atmosphere. Almost all steel structures are coated with
some protective materials. Other metals such as copper, nickel, and
titanium, appear to be resistant to pollutants. Pollutants may,
nonetheless, have important effects on materials. For instance,
pollutants may affect other characteristics of metals such as their
bending strength (see Haynie [21]).

Sulfur products and particulates are also suspected of damaging
fabrics, carbonate stone, and electrical equipment. A study by
Brysson [9] suggests that sulfur products reduce the breaking strength
of cotton fabrics. The exposures he used, unfortunately, are two orders
of magnitude above normal. These exposures did detectable damage to
fabrics. The cotton fabrics appear to lose about .025% of their breaking
strength every ten years per $\mu g/m^3$ of sulfur dioxide. Needless to say,
at normal exposures, sulfur would have only a small effect on the life of
most cotton fabrics. Gauri and Sarma [18] have shown that sulfur dioxide
(in concentrations five orders of magnitude above normal) reacts with
marble and limestone. The actual damage from these reactions is not
quantified but presumably the pollutants can eventually deface statues
and building. Campbell [12] has shown that particulates affect brass
plated electrical equipment by forming a surface film. Accurate figures
on the extent of these damages are unlikely because individual
businessmen are rarely in a position to know what effect an incremental

unit of a particular pollutant has upon the corrosion rate of his
materials. Nonetheless, several surveys yield estimates of the national
annual damages from material exposures to pollutants.[4] Assuming these
effects are equally divided between sulfate and particulates and
proportional in a local area to total exposures, sulfate causes about
$.012 and particulates cause about $.002 per person per $\mu g/m^3$ annually.

The corrosion of paint by pollutants is documented by both
laboratory and unplanned natural experiments. Spence [35] examines the
effect that sulfur dioxide, nitrogen oxide, and ozone have upon paints in
exposure chambers. The excess corrosion rate caused by 1 $\mu g/m^3$ of ozone
is about 5×10^{-6}mm/year. The effect of sulfur dioxide, especially in
conjunction with humidity, is larger than the ozone effect. Other
measures of the effect of air pollution on paint come from surveys of
household maintenance activities. Michelsin and Tourin [35] compare five
cities and Booz Allen and Hamilton [8] compare four sites within
Philadelphia. The former study finds a distinct relationship between the
frequency of exterior painting and particulates while the latter study
finds no relationship. Both studies neglect to control for the income of
the owner and the value of his house, both of which affect the frequency
of maintenance expenditures (see Mendelsohn [33]). This problem is
probably more serious with the Booz Allen Hamilton study because of the
association within an SMSA of poor air quality and low income
neighborhoods. Another problem with the method of both studies is that
part of the cost of additional particulates may be borne by homeowners by
living in more poorly painted dwellings or by buying more expensive
pollution resistant paint. The frequency of expenditures on exterior
painting underestimates the total burden of the pollutant. Using the

Michelson and Tourin study, 1 μg/m^3 of particulates increases the frequency of painting .005 times a year. Since this increases painting about 1%, approximately 1% of annual exterior painting (about 10 million dollars) is related to each μg/m^3 of particulates.[5] This amounts to about $.04 per person per μg/m^3 of particulates. The range of possible values for this parameter are between $.002 and $.20 per person per μg/m^3. Part of paint losses are quite likely a result of sulfate. This effect is probably also about $.04 per person with the same range of possible values. A large proportion of the material damages caused by pollution is due to paint corrosion.

Oxidants and nitrogen oxides are accused of causing fading in dyes used on fabrics. Fading from nitrogen dioxide results in reddening while oxidant fading results in a bleached, washed-out appearance (see EPA [49]). Salvin [32], using gas chambers, has demonstrated fading from NO_x at concentrations of 38,000 μg/m^3. Manufacturers have since shifted to more expensive dyes which resist this effect (see EPA [49]). Upham [51] tests the effect of nitrogen dioxide on cellulose fabrics and finds that the useful life of such fabrics is reduced because of fading by .5% per μg/m^3 of nitrogen dioxide per year. Assuming that about $1 billion of new clothing can potentially be damaged by nitrogen dioxide each year, the annual national damages of 1 μg/m^3 of nitrogen oxide is about five million dollars. This is equivalent to about $.02 per person per μg/m^3 of nitrogen oxide. The range of this parameter probably lies between $.0008 and $.50 per person per μg/m^3.

Table V-1 summarizes the expected dose responses for each pollutant upon materials. Estimated standard deviations for each parameter are also shown. Assuming linear dose response curves and equal proportional

Table V-1

Annual Rate of Damage to Materials from Four Air Pollutants[a]

Type of Material	Sulfur Oxides	Nitrogen Oxides	Particulates	Oxidants[b]
Electrical Equipment	30 (20)	0 (.5)	1 (3)	0 (.5)
Other Metals	1 (5)	0 (.5)	.5 (8)	0 (.5)
Paint	100 (200)	0 (.5)	10 (50)	2.5 (50)
Dyes	0 (.5)	28 (150)	0 (.5)	10 (10)
Carbonate Stone	12 (10)	0 (.5)	2.5 (3)	0 (.5)
Others	50 (50)	14 (20)	.5 (2)	100 (100)
Total[c]	193 (207)	42 (151)	19 (51)	135 (112)
Expected National Damage (10^6/year)	500	520	332	934

[a] The damages are expressed in terms of mills lost per person per $\mu g/m^3$ except as noted.

[b] Oxidants are formed from nitrogen oxide emissions.

[c] Estimate of variance for total assumes statistical independence amongst the types of materials.

effects across people, the expected values shown in Table V-1 imply national damages of about 2.2 billion dollars for materials. These national figures are consistent with estimates developed by Waddell [5]. Note that the range of uncertainty around these estimates is quite broad--they are not significantly different from zero. Additional experimental evidence could narrow the uncertainty of these dose response curves significantly.

C. Damage to Vegetation

There are a number of studies in recent years which have demonstrated plant damage from sulfur oxides, nitrogen oxides, acid rain, and oxidants. Resistance to each pollutant varies tremendously across species. More studies of the effects on different plants are clearly needed. Another problem is that appropriate measurements of plant damage are not always available. For instance, how pollution impacts the yield from crops (both the quality and the quantity per acre), damages the appearance of ornamental plants (altered leaves or growth), and reduces the pleasure of recreational ecosystems (stunted growth, wild plant deaths, or altered system equilibriums) is not always evident from reported results. Other measures such as the rate of photosynthesis or even leaf injury do provide sensitive qualitative evidence of plant damage, but at least with most crops, such measurements do not provide sound estimates of the extent of economic damage. Though these physically sensitive measurement techniques may identify important qualitative relationships, more relevant measures need to be used in order to make quantitative estimates of the extent of pollution-caused plant damage.

Sulfur dioxide concentrations correlate with leaf injury in a number of experiments on pines, poplars, tobacco, desert plants, and ornamental trees. Exposures in the range of 4-8,000 $\mu g/m^3$ are noticeably harmful and the damage increases rapidly with the length of the exposure. Lower plant types, such as lichen and moss, appear to be especially sensitive to sulfur dioxide. Concentrations as low as 200 $\mu g/m^3$ cause serious damage and death in these lower plant types. Except for the probable damage to lichens and moss, there is little direct evidence that typical long term exposures of sulfur dioxide have discernable effects on plant life. The effect of 1 $\mu g/m^3$ of sulfur dioxide on crops is assumed to reduce plant value by .05% though it could range between .005% and .5% per $\mu g/m^3$. The additional effect on ornamental plants is probably almost equal to that on crops. The damage from additional sulfur dioxide on ecosystems is probably not noticeable to the average user. Only with unusually high exposures would the effects be visible. Damage to natural habitats is thus a possible but most likely small effect.

Oxidants, including ozone and photochemical smog, appear to have severe impacts on agircultural crops. Alfalfa, broccoli, corn, cotton, cucumber, lettuce, oats, potatoes, radishes, sorghum, spinach, tobacco, tomatoes, and citrus trees have been noticeably damaged when exposed to concentrations of 10,000 $\mu g/m^3$ or higher of oxidants for a few hours. In addition to these acute effects, there have been a few studies of long term effects from exposures to "smog". Cotton grown in existing levels of California smog has a lower yield of between 10 and 20% than equivalent plants grown in filtered air. Similar studies of spinach, radishes, and lettuce demonstrate that the yield could be reduced by more than

one-third if the plant is exposed to the highest levels of existing smog.
Costumes and Sinclair [14] and Miller [36] both suggest that low level
chronic exposures to oxidants cause serious leaf damage and susceptibility
to insect attack. Given this evidence, it is assumed that an additional
$\mu g/m^3$ exposure to oxidants reduces plant yield by about .5%. The
probable range of this parameter is between .05% and 5% of total yield.
Oxidants appear to be responsible for most of the damage to plants
caused by air pollution.

Nitrogen dioxide is also harmful to some plants. Laboratory
experiments have detected subtle effects. The exact nature of the
effects in natural settings is difficult to ascertain because of the
close atmospheric interaction between nitrogen oxides and oxidant
formation. Air chamber experiments reveal that concentrations of 620
$\mu g/m^3$ administered for 10 to 19 days result in observable damage to pinto
beans. Exposures of nitrogen dioxide of between 310 and 1000 $\mu g/m^3$
result in lower growth rates of between 15 and 25% in tomatoes. The
effect of 1 $\mu g/m^3$ of nitrogen oxide probably reduces the value of
susceptible plants by about .05%. The range of this estimate lies
between .005 and .5% per $\mu g/m^3$ of nitrogen oxide. There is no evidence
to suggest that wild plants are susceptible to nitrogen oxides and so
damage to ecosystems is probably negligible. There is also no evidence
to suggest that nitrogen oxides have appreciable impacts on ornamental
plants.

Table V-2

Annual Effect of Additional 1000 Tons of Emissions on Vegetation[a]

(dollars of damage-standard deviation in parenthesis)

	Sulfur Oxide	Nitrogen Oxide	Oxidants[b]
Crops	1000 (3000)	500 (1500)	10000 (15000)
Ornamental	1000 (3000)	500 (1500)	5000 (8000)
Natural Habitat	100 (600)	50 (300)	500 (1500)
Total	2100 (4284)	1050 (2142)	15500 (17000)

[a] The location of the plant is assumed to be independent of vegetation damage because vegetation is distributed nearly uniformly. Damages are, therefore, expressed in terms of emissions.

[b] Oxidants form in the atmosphere from nitrogen oxides and hydrocarbons. Power plants contribute to oxidant formation through emissions of the chemical inputs. Oxidant damage should, therefore, be allocated to these other chemicals in proportion to their contribution to oxidant exposures to vegetation.

D. Chronic Health Effects

The health consequences of long term low level exposures to air
pollution are among the most difficult dose response effects to study and
probably the most important. In order to examine these relationships
properly, one needs large populations, controlled environments, and a
long period of time. The expense and moral dilemma of intentional human
exposures to harmful pollutants thwart the use of planned experimentation.
Greater reliance, therefore, must be placed on retrospective unplanned
studies. Even though difficult to study, it is nevertheless clear that
the magnitudes of effects from long term dosages are overwhelming when
compared to acute effects. According to Mendelsohn and Orcutt [34], over
90,000 lives are lost each year in the United States as a result of
persistent nonzero air pollution. This is equivalent to losing about
one half billion days of expected lifetimes each year. Acute health
effects, as opposed to chronic effects, are associated with less than
one million days lost each year. The chronic effects of air pollution
on health may well be the most critical set of dose response curves in
the entire model. Because of the importance of these health effects, the
evidence behind them is discussed in greater detail than the previous
dose response functions.

The chronic effects of long term exposures to pollutants are large
but they are also quite subtle. The calculated 90,000 deaths from air
pollution each year are only 5% of all deaths. Since the mortality rate
in the country is about one in a hundred, only one in every two thousand
persons may die from air pollution. If one collected a survey of ten
thousand people, a relatively large sample, one could only expect about

five deaths from air pollution. In order to detect such a rare event
as excess deaths, one needs substantial numbers of observations of
individuals. Another problem is that other factors may also effect the
mortality rate and overpower the effects of air pollution. If the study
does not adjust for these other factors, the results would be less
accurate and could possibly be completely misleading. The second
requirement of a good study, then, is that unwanted influences be taken
into consideration. This generally requires one to collect, or at least
be aware of, extensive information about each individual. A third
requirement of a careful study is that the variables of interest, in
this case, air pollutants, vary across the sample space so that one has
at least a chance to detect the consequences of different levels of
treatment. Classical epidemiological studies, constrained by budget and
time limitations, generally have not satisfactorily met these three data
requirements. Either the samples are too small, information about each
observation is insufficient, or independent variations of pollutant
levels are inadequate. Economists, aided by their historical experience
studying unplanned environments, have made important contributions to
this field by bringing new data to bear on these analyses. However,
additional work is urgently needed with yet better data in order to
improve upon our insights into the nature and magnitude of these health
benefits.

There are two broad classes of unwanted variation which should be
accounted for: personal and area wide factors. Personal factors,
especially age, can cause wide variations in mortality rates if the
personal factors differ across areas. Because of the importance of these
factors, information about only the people who die or who are ill is

misleading by itself. Differences in who dies across areas can be solely a function of the characteristics of the population in those areas. The environments or, more specifically, the levels of pollution may have no effect on health. Most epidemiological studies, consequently, are carefully sampled in order to reduce the impact of at least three major personal characteristics: age, race, and sex. Other variables such as marital status, income, and education are occasionally proxied with a single index or are completely ignored. The absence of these other personal characteristics in most studies is probably causing some errors but the magnitude of these errors is probably small.[6]

One of the most important problems with almost all classical epidemiological work is the failure to properly account for area wide variables. Faced with the enormous cost of collecting extensive data in a number of sites, epidemiologists have retreated into studying single metropolitan areas. The implicit assumption in this sampling method is that area wide variables are approximately the same throughout the metropolitan area. After carefully choosing sample populations to avoid variations in personal characteristics, the epidemiologists generally assume that the remaining variation can be freely ascribed to differing pollutant levels. At least in the United States, though, there are systematic relationships within single metropolitan areas which tie many area wide variables to levels of pollution. Areas with the highest levels of at least primary pollutants are usually adjacent to industrial areas, downtown, old, and deteriorating. There are a number of potential reasons why these areas would have higher mortality rates in addition to just air pollution. Unless an effort is made to capture these other characteristics, it is probably misleading to ascribe different levels of health among these areas solely to air pollution.

Another limitation of almost all epidemiological studies is the stark absence of quantitative estimates. Of over 200 studies reviewed in Lave and Seskin [31], only a handful provide an estimate of the size of the health response to the pollution stimulus. Only a few additional studies report findings in a format whereby one can infer the possible magnitude of the effects. Almost all of the studies report only simple correlations between a health effect and a population index. Though this literature is suggestive, it provides little information about the relative importance one should place on specific pollutants.

Lave and Seskin [29, 31] compare over one hundred cities and make an effort to include at least a few area wide characteristics. Only their most recent studies, however, adequately account for variations in personal characteristics. Their studies suffer somewhat from the large unit of observation, a metropolitan area, but better data is not readily available. These studies, nonetheless, make an important contribution by encouraging examination of relationships across many sites.

Mendelsohn and Orcutt [34] contribute to the understanding of chronic mortality by bringing to bear large quantities of data in their analysis. Their study merges the extensive information collected by the Public Use Sample with 1970 death certificates by 24 age, race, and sex categories and by county groups. Altogether, two million death certificates and two million Census returns are utilized in the final data set. Not only is this the largest data set yet used for pollution mortality studies, but it also contains more information about individual cells and their area wide characteristics than previous studies. Information recently collected by EPA on sulfate, sulfur dioxide,

nitrate, nitrogen dioxide, carbon monoxide, particulates, and ozone
concentrations is merged with the county group information. The data
set contains measurements of more pollutants than past studies. This
massive data set offers an improved opportunity to separate out the
effects of individual pollutants.

In a series of regressions, Mendelsohn and Orcutt demonstrate the
correlation between area wide characteristics and primary pollutants.
Including area wide characteristics in the regression equations reduces
the size of primary pollutant coefficients. Studies which fail to
include area wide variables probably overestimate the effect of primary
pollutants on health. Personal variables, especially age and sex, are
also shown to be important: the coefficients of all the independent
variables across age and sex cells appear to be significantly different.
Age and sex not only directly effect the level of mortality, but they
also exert influence through different responses to the independent
variables.

The mortality dose response curves in Mendelsohn and Orcutt are
shown in Table V-3. Men are more sensitive to higher pollution than
women and the size of the pollution coefficients steadily increases with
age. People under 18 do not suffer any observable effects from pollution
Another finding is that the dose response curves are approximately linear
within the range of ambient doses present in the United States. Since
the study includes a wide range of pollution concentrations, it is
reasonable to assume a linear dose response from plausible exposures to
pollutants. Interactions among pollutants were also not detectable in
the study. Additional units of harmful pollutants may have the same
effect regardless of the background air quality.

Table V-3

Predicted Mortality Responses to Air Pollution Levels[a]

	Sulfate	Nitrate	Particulates	Carbon Monoxide[b]	Sulfur Dioxide	Nitrogen Dioxide
Males						
18-24	20 (40)	- 18 (33)	3 (24)	-1 (3)	-.4 (3)	3 (2)
25-44	20 (4)	- 23 (12)	.3 (.8)	25 (8)	2 (1)	2 (.7)
45-64	167 (20)	-106 (67)	-10 (4)	135 (46)	15 (5)	-6 (4)
65 +	822 (103)	-215 (336)	6 (20)	222 (243)	26 (26)	-80 (20)
Female						
18-24	14 (2)	3 (7)	.5 (.4)	14 (5)	1 (.6)	-.4 (.5)
25-44	9 (3)	7 (8)	.1 (.9)	16 (6)	1 (.7)	.2 (.6)
45-64	80 (10)	-37 (34)	-2 (2)	85 (23)	13 (3)	-5 (2)
65 +	569 (66)	383 (214)	5 (13)	126 (152)	27 (17)	-52 (12)

a Source: Mendelsohn and Orcutt [34]. The responses are measured in terms of 10^{-6} deaths per $\mu g/m^3$ person year.

b These figures are in terms of 10^{-6} deaths per mg/m^3 person year for carbon monoxide.

Among the tested pollutants in the Mendelsohn and Orcutt study, the most toxic pollutant appears to be sulfate. Sulfate not only has relatively large positive coefficients, but it also has a consistent significant effect across all adults. Sulfur dioxide and carbon monoxide may also be harmful but they are not of the size or significance of sulfate. Nitrogen dioxide and ozone frequently have negative coefficients which imply beneficial effects. Though it is unlikely that either pollutant is, in fact, good for one's health, it is even more unlikely that they are very harmful.

Since any one study may be subject to a peculiar error, it is helpful to compare these results with other epidemiological studies. Shy [44], for instance, suggests that nitrogen dioxide has large impacts on the morbidity rate of children. Shy's finding, however, has been attacked by Warner and Stevens [53] because large quantities of nitrate and nitric acid are present in the more polluted communities of the Shy study. These other pollutants are more likely than nitrogen dioxide to be the cause of the high illness rates observed in these communities. The results of the Lave and Seskin [31] studies are shown in Table V-4. Only two pollutants are included in these studies and their minimum values instead of their mean values are sometimes used. Despite these differences, the pollutant coefficients for suspended sulfate and particulates are within an order of magnitude of the corresponding coefficients in Table V-3. Winkelstein's results [54] for elderly men suggest a dose response curve for sulfates of about 3×10^{-4} deaths per $\mu g/m^3$. Sprey and Takacs [46], looking at white males over 65 generate a dose response curve of about 2.5×10^{-4} deaths from heart disease per unit of sulfate. Given the proportion of elderly

Table V-4

The Effect of Air Pollution on Urban Mortality[a]

1959

White
Male

	Minimum Sulfate	Mean Particulate
15-44	.04 (1)	7 (7)
45-64	4.4 (5.2)	54 (38)
65 +	45.3 (20)	169 (150)

Female

15-44	-.25 (.54)	10 (4)
45-64	8.1 (3)	60 (22)
65 +	66.4 (18)	71 (130)

1969

White
Male

	Minimum Sulfate	Mean Particulate
15-44	-3.42 (1.5)	34.3 (10)
45-64	3.48 (7.7)	132.2 (51)
65 +	66.4 (30)	268 (200)

Female

15-44	-1.73 (.9)	9.4 (6)
45-64	2.05 (3.5)	37.9 (23)
65 +	52.1 (21)	354 (140)

[a] Source: Lave and Seskin [31]. The figures are 10^{-6} annual deaths per $\mu g/m^3$ person year.

who dies from heart disease, the deaths from all causes may lie between 4 and 6 x 10^{-4} per $\mu g/m^3$ of sulfate. The range of values in Mendelsohn and Orcutt [34] for the equivalent age, race, and sex cell is between 3.7 and 8.7 x 10^{-4} deaths per $\mu g/m^3$ of sulfate depending upon the confounding variables included in the regression. The results across several studies are reasonably consistent and support the findings in Mendelsohn and Orcutt.

Toxicological studies of long term exposures also tend to reinforce the qualitative findings in Mendelsohn and Orcutt [34]. Man appears to be very resistant to nitrogen oxides. Humans exposed to 2000 $\mu g/m^3$ of nitrogen oxide for 180 days show only minor biochemical changes (see Kosmider [28]). Though nitrogen oxide exposures at high concentrations may be harmful, both toxicological and epidemiological evidence suggest that exposures at normal levels (100 $\mu g/m^3$) do not have noticeable effects. The epidemiological evidence that sulfate is far more hazardous than sulfur dioxide is also supported by laboratory animal experiments. Continuous exposures to guinea pigs of 10,000 $\mu g/m^3$ of sulfur dioxide have no effects.[7] In contrast, exposures of as little as 70 $\mu g/m^3$ of sulfates result in easily detectable consequences to the lungs of guinea pigs. Several toxicological studies provide further supportive evidence of the qualitative hypothesis that sulfates are more hazardous than sulfur dioxide.

Though not emitted in large quantities, trace substances may have substantial impacts on both human health and ecological systems. Lead, mercury, and selenium are all emitted in coal fired power plant plumes and they are suspected to be dangerous even in small quantities.[8] Mercury poisoning, for instance, is evident in workers exposed to just

$30 \ \mu g/m^3$ of mercury in the air. Thus, even though these pollutants are present only in small concentrations, they may still be quite dangerous. There is not enough evidence, unfortunately, to make reasonable estimates of the dose responses of trace metals on humans. Since they were not explicitly included in the statistical analyses of the effects of the other pollutants, it is possible that one of the included pollutants, possibly particulates, is acting as a proxy for toxic trace elements. This would be fortunate, if true, because abatement techniques applicable to at least small particles also reduce trace elements. The model would, under these circumstances, accurately predict the consequences of engaging in particulate control. The absence of explicit accounting of trace metals does not automatically bias the results or damage the implications of this analysis. With further research into the health effects of trace elements, hopefully, the toxic effects of trace elements can be analyzed more precisely.

Because statistics on morbidity are rarely published and are expensive to collect, there are only a limited number of epidemiological studies of the effects of air pollution on rates of illness. Since the respiratory system is the most obvious target of air pollution, initial studies concentrated on measuring the effect of air pollution on respiratory illnesses. Recent studies of deaths from air pollution suggest that other bodily systems may also be affected by poor air quality. In particular, deaths from circulatory system failure occur more frequently in highly polluted areas. Though this cause of death may just be an indirect consequence of respiratory stress, the impact of air pollution on circulatory illness may, nevertheless, be an important effect. Current studies, though, focus almost entirely upon respiratory diseases, so it is impossible to estimate the magnitude of nonrespiratory morbidity due to pollution.

In contrast to mortality rates where death is a clearly measured event, morbidity rates can be misleading because not all doctors have the same concept of the exact symptoms of each disease. For example, any persistent cough, only a persistent cough with phlegm, or only a persistent cough with both phlegm and shortness of breath all may be judged by different doctors as bronchitis. Past studies, therefore, are often careful to define the diseases they are measuring (the problem is even more serious with self reporting surveys). Even if reporting is consistent, however, there is nonetheless a conceptual problem of how to describe symptoms of a continuous nature with a limited set of disease categories. This is particularly a problem with measuring chronic illnesses. One method of handling the problem is to establish distinct categories for each class of symptoms such as in Table V-6.

In addition to chronic illnesses, there are a number of acute respiratory illnesses which might be associated with air pollution: croup, acute lower respiratory illness, bronchitis, and pneumonia. Whereas chronic symptoms are prevalent in adults, acute illnesses are concentrated in both the very young and the very old.

The primary source of information about long term effects of pollution on morbidity rates come from the CHESS studies, USEPA [50]. The CHESS studies are a series of analyses of acute and chronic respiratory illness rates within four metropolitan areas. In general, the chronic disease studies utilize the parents of school age children and the acute studies utilize the children themselves. Three or four sites within each metropolitan area are compared. Age, race, sex,

Table V-5

Measures of Chronic Lung Disease

1) No symptoms
2) Persistent cough for < 3 months
3) Phlegm for < 3 months
4) Cough alone for > 3 months
5) Phlegm alone for > 3 months
6) Cough and phlegm for > 3 months
7) Cough and phlegm for > 3 months
   and shortness of breath.

smoking status, and sometimes occupational exposures are available for each individual. Socioeconomic status and area wide variables are assumed to be captured by an all-encompassing index or are assumed to be constant across all the sites. Each metropolitan area is analyzed individually in the CHESS documents.

Berman [6] combines all the CHESS sites along with a few other locations in order to analyze the contribution of various pollutants to bronchitis rates (persistent cough with both phlegm and shortness of breath) and mean symptom scores. He finds that sulfur dioxide and particulates do not have a significant impact on either bronchitis rates or mean symptom scores. Sulfate, on the other hand, appears to increase both mean symptom scores and bronchitis rates. Because the mean symptom score is an awkward average, it is difficult to interpret these results. The reported effect on bronchitis, however, suggests that a 1 $\mu g/m^3$ increase of sulfate will increase the bronchitis rate by a little less than one-half percent. Sulfate exposure increases the rate of bronchitis approximately 50 times more than mortality rates. Morbidity dose response estimates, however, are probably far less certain than mortality estimates because of the relative paucity of morbidity statistics.

In addition to these chronic effects, the CHESS study also examines acute respiratory effects in children. Sulfur dioxide and sulfate are correlated with croup, lower respiratory illness, and bronchitis in the Utah data. A comparison across the Rocky Mountain sites provides no compelling evidence of an association between pollution and any of these morbidity measures. The New York sites fail to provide any positive correlations between pollution and acute respiratory ailments. Pneumonia, another chronic disease, is not correlated with air quality at any site. The sample provides further contradictory evidence when it is divided between long term residents and children who moved there in the last three years. Areas with good air quality were consistently worse for children who recently moved there. Given the conflicting evidence presented in CHESS, it is difficult to know whether air pollution has any effect on acute respiratory illness. Table V-6 presents the expected effects as well as the uncertainty of child morbidity responses to several air pollutants.

E. Acute Health Effects

Acute health effects refer to symptoms which are caused by short term exposures. These exposures typically vary from a few hours to a few days. Acute, unlike chronic, effects have received a great deal of public attention. Unusual incidents of sharply increased pollution levels over metropolitan areas have sometimes resulted in dramatic increases in hospital admissions and deaths. These well publicized events have encouraged many studies of the relationship between daily air quality and daily mortality and morbidity rates (see, for example, Buechley [10,11], Hextor and Goldsmith [25], Hodgson [26], Glasser and Greenburg [19], or Lave and Seskin [30]). Because acute experimental doses can be administered quickly and inexpensively, many toxicological experiments have also focused on acute effects. Given that the expected magnitude of acute effects is small in comparison to chronic effects, a disproportionate share of all health effect studies focus on acute phenomenon.

The most dramatic acute effects are probably perturbations in daily

Table V-6

Child Morbidity from Chronic Exposures to Air Pollution[a]

(10^{-5} x cases per $\mu g/m^3$ person year)

	Sulfur Dioxide	Sulfate	Particulates
Bronchities Rates[b]	-.9 (6)	447 (250)	36.0 (35)
Lower Respiratory Disease[c]	7.5 (10)	79.2 (60)	-
Croup[c]	7.5 (10)	79.2 (60)	-
Pneumonia[c]	1.3 (5)	50 (100)	-

[a] Standardized errors in parenthesis.

[b] Figures obtained from Table 2 in Berman [6].

[c] Figures extrapolated from USEPA [50].

mortality rates which are associated with fluctuations in air quality.
Several studies find that changes in sulfur dioxide, smoke shade, or
carbon monoxide concentrations are correlated with daily mortality. In
four of five cities, Lave and Seskin [30] are unable to find an
intertemporal correlation between air pollution and daily mortality.
In Chicago, however, Lave and Seskin are able to find positive associations
between daily deaths and nitrogen dioxide and sulfur dioxide. The
magnitude of these associations implies that 1 $\mu g/m^3$ of sulfur dioxide
would cause about 1.5×10^{-8} deaths per day. Alternative specifications
of the lagged relationship between these pollutants and death indicate
that these associations are somewhat weak. The evidence across
specifications and across metropolitan areas does not reveal a strong
association between air quality and daily mortality. Additional evidence
on the acute effects of carbon monoxide have been gathered through a
number of experiments (see Binder and Feinberg [7a]). Acute mortality
from carbon monoxide poisoning is not expected to occur below about
750 mg/m^3 for 10 hours. Since exposures above 50 mg/m^3 are rare, carbon
monoxide probably has only a negligible effect on daily mortality rates.

Even if one were convinced that unusually high temporary exposures
to pollutants caused immediate deaths, the significance of these deaths
is still an important question. How many days of expected lifetime is
lost with each of these deaths? Lave and Seskin assert that the expected
lifetime lost from acute effects is at least ten days. Other evidence
seems to suggest that it may not be much more than about 10 days on
average. Studies of observed critical episodes note that many of the
people who die during these episodes are already critically ill (see
Logan [32]). Mendelsohn and Orcutt [34] could find no consistent or

significant relationship between daily fluctuations of pollution and
annual death rates.[9] The expected lifetime lost from acute effects is
small enough not to effect the annual mortality rates. Expected days
lost per death from acute exposures to pollutants may lie anywhere
between 0 and 360 days.

Toxicological experiments on man and other animals reinforce the
epidemiological findings that normal acute exposures will not
substantially alter mortality rates. Enormous exposures must be
administered to even susceptible species in order to induce detectably
higher mortality rates. The fifty percent lethal dosage for rabbits and
squirrels to nitrogen oxide concentrations is 50,000 $\mu g/m^3$ for 3 days
and 80,000 $\mu g/m^3$ for 2 hours, respectively. Man is suspected to be more
resistant to nitrogen oxides than either of these species and he is
rarely subjected to doses exceeding 300 $\mu g/m^3$. Exposures to guinea pigs
(whose resistance is similar to man's) of up to 20,000 $\mu g/m^3$ of nitrogen
oxide have no observable effect on mortality rates. It is not surprising
that nitrogen oxides are not observed to increase daily mortality rates.

Sulfur dioxide exposures of mice, rats, dogs, and guinea pigs
suggest that acute exposures of between 18,000 to 500,000 $\mu g/m^3$ do not
effect mortality rates. The toxicological findings provide further
evidence that temporary exposures to sulfur dioxide are not very harmful.
Sulfate and sulfuric acid mist exposures to guinea pigs, on the other
hand, suggest that either of these sulfur products could potentially have
serious acute effects. Because historical data on sulfates are generally
unavailable, epidemiological studies have not tested the potency of
sulfates as opposed to sulfur dioxide. It is possible that the sulfur
dioxide coefficients in the daily mortality studies are actually serving

as proxies for correlated fluctuations in the level of sulfate. If one attributes to sulfate the sulfur dioxide effect in the Lave and Seskin study, 1 µg/m^3 of sulfate per day would approximately cause about 9 x 10^{-7} days of expected lifetime lost per person. A population of about one million persons exposed to the national average pollution level would lose about 3000 days from acute deaths per year. This is rather small in comparison to the two million days expected to be lost from chronic deaths per year due to the same level of exposure.

Epidemiologists have also investigated the impact of acute exposures on morbidity rates.[10] Daily particulates, smoke shade, and sulfur dioxide exposures are related to hospital admissions, length of stay, and asthma attacks. Nitrogen oxides are correlated with asthma attacks. Carbon monoxide exposures for 10 hours at 182 mg/m^3 may cause mild headaches and higher exposures can result in dizziness and nausea. Time series studies of oxidants suggest that photochemical oxidants cause symptoms including eye irritation, headache, and chest discomfort. Almost all the pollutants may have some acute effect on morbidity.

The acute dose response of headaches to carbon monoxide concentrations is approximately 1 x 10^{-3} per ten hour exposure of mg/m^3 per person (see Binder [7]). At this rate, acute exposures to the national average carbon monoxide readings would result in about one mild headache per person per year.

Hammer [20], in a Los Angeles study of nurses, finds that oxidants are associated with headache, chest discomfort, and eye irritation. An approximate dose response curve for each of these minor effects is: 3 x 10^{-4} for eye irritation, 1 x 10^{-4} for headache, and 2 x 10^{-4} for chest discomfort per µg/m^3 of oxidant per day. With an average concentration of about 60 $\mu g/m^3$ of oxidants, the average person would experience between 2 and 7 days a year with symptoms caused by oxidants alone.

There are two time series analyses of asthma attacks in CHESS from Utah and New York. Both studies compare daily air quality with the frequency of asthma attacks in people known to suffer from asthma. Since such people make up about 4% of the total population, the dose response of the population towards asthma is assumed to be 1/25th of the recorded dose rates for this susceptible sample. Sulfur dioxide, particulates, and sulfate are all positively correlated with the frequency of daily asthma attacks. The sulfate effect, however, is noticeably larger than the effects caused by the other pollutants as can be seen in Table V-7.

Acute exposures to air pollutants produce a wide range of physiological changes in animals. Though measures such as increased pulmonary flow resistance and lung surfactant activity are not indices of respiratory disease, they do reflect physiological changes which may be the precursors of respiratory failure or inflammation. Several animal studies have provided important insights about the harmful agents in air pollution. Amdur [2, 3] in a series of experiments is the first researcher to demonstrate that sulfates and sulfuric acid mist have significant impacts on lung function while sulfur dioxide is relatively harmless. This important finding has led to definite improvements in the interpretation of epidemiological studies where it is now widely accepted that sulfur dioxide serves as a proxy for sulfate when the latter is omitted. Amdur also demonstrates that guinea pig lung function is much more sensitive to submicron particles than to larger more massive particles. Amdur has shown that certain sulfates may be more hazardous

Table V-7

Health Effects Due to Acute Exposures to Air Pollutants[a]

	Oxidants	Sulfur Dioxide	Particulates	Sulfate	Carbon Monoxide
Headache	100 (50)	-	-	-	1 (2)
Chest Pain	200 (100)	-	-	-	-
Eye Irritation	300 (150)	-	-	-	-
Asthma	-	4 (8)	7 (2)	37 (45)	-
Premature Death					
18-46	-	.1 (1)	.01 (.1)	1 (10)	-
65 +	-	.3 (3)	.1 (1)	7 (70)	-

a   The health response is measured in terms of 10^{-6} days/$\mu g/m^3$ person day or 365×10^{-6} days per $\mu g/m^3$ person year. Standard errors are in parenthesis.

than others (see Table V-8). Unfortunately, little is now known about the formation and creation of different types of sulfates so that this evidence is difficult to incorporate in the model.

Brief exposures to nitrogen dioxide at high concentrations (2000 $\mu g/m^3$ and higher) result in slight physiological changes in the lungs of most animals. Short exposures to humans of between 2000 and 10,000 $\mu g/m^3$ result in temporary increases in airway resistance. Exposures below these levels are not evident in the literature. Brief nitrogen oxide exposures of around 200 $\mu g/m^3$ (the highest concentration in many areas) could well have some physiological impact on a population but it is probably relatively mild and temporary.

There are indications that pollutants can produce physiological changes in lungs after even brief exposures. However, with exposure levels no greater than an order of magnitude above normal ambient concentrations, only relatively mild symptoms can be detected. Furthermore, there is considerable evidence that lung function returns to normal relatively quickly after the exposure. The magnitude of the final effect of temporary exposures to air pollutants appears to be small. Nonetheless, acute exposures to air pollutants do have annoying effects and people who are already ill can be made more uncomfortable by even temporary poor air quality. Acute effects, though minor, are included in this modelling effort.

Table V-8

Ranking of Sulfates on Basis of Acute Irritant Potency[a]

% Increase Resistance per $\mu g/m^3$ of XSO_4

H_2SO_4	.410
$Zn (NH_4)_2 (SO_4)_2$	.135
$Fe_2 (SO_4)_3$	.106
$Zn SO_4$	.079
$(NH_4)_2 SO_4$	.038
$NH_4 SO_4$	.013
$Cu SO_4$	.009
$Fe SO_4$	.003
$Mn SO_4$	-.004

[a] Source: Amdur and Corn [3]. Abnormally high
acute exposures of each sulfate are given to
normal guinea pigs and respiratory flow
resistance is measured immediately after the
exposure. Whether these rankings are
appropriate for low level chronic exposures
to man is not known.

Footnotes

1 See Orcutt and Glazer's "Research Strategy, Microanalytic Modeling, and Simulation." ISPS Working Paper 793, Yale University, 1977.

2 Three studies which report such results are Charlson, et al.[13], Horvath and Nath [27], and Ellinger and Royer [17]. For more information about the effect of particulates on visibility, one should consult these studies.

3 The National Academy of Sciences [38], p. 624, somewhat arbitrarily estimates that acid rain may cause $500 million of damage. This figure includes sulfur damage to buildings and materials which we include separately. The corrected figure for acid rain should be about $400 million nationally. The true number probably lies between $40 and 1600 million.

4 These surveys of material damages are reviewed in Waddell [52]. They form the basis of Waddell's national damage estimates.

5 These estimates were obtained from Residential Alterations and Repairs, United States Commerce Publication C50-74, U.S. Government Printing Office, Washington, D.C., 1976.

6 The inclusion of proxies for personal characteristics in the Mendelsohn and Orcutt [34] study fail to alter the significance or magnitude of the pollution coefficients. Once age, race, and sex are controlled, further personal variables may no longer effect or bias the pollution coefficients.

7 Experimental results by Alarie [1] and Rylander [41] show no effects on guinea pigs given exposures of sulfur dioxide up to 10,000 $\mu g/m^3$ for one year and 20,000 $\mu g/m^3$ for 20 days, respectively.

8 Two useful reviews of the effects of trace substances are: "Health Effects of Mercury," Science Applications, Inc., EPRI EC-224, 1976, and "Health Effects of Selenium," Science Applications, Inc., EPRI-571-1, 1976.

9 Controlling for unwanted variations from extraneous sources and the mean level of air pollution, Mendelsohn and Orcutt [34] find that variations in measured ambient concentrations around that mean do not contribute to annual death rates. Areas that had both high and low exposures had the same death rates as areas which had just mean exposures.

10 For a review of this literature, see the Appendix of Lave and Seskin [31], EPRI [15], and National Academy of Sciences [38, 39].

Bibliography

1. Alarie, Y.C., et al. "Long Term Continuous Exposure of Guinea Pigs to Sulfur Dioxide." Archives of Environmental Health, 21 (1970), 769–777.

2. Amdur, M.O. "The Influence of Aerosols Upon the Respiratory Response of Guinea Pigs to Sulfur Dioxide." American Industrial Hygiene Quarterly, 18 (1957), 149–155.

3. _____, et al. "Respiratory Response of Guinea Pigs to Sulfuric Acid and Sulfate Salts." Symposium on Sulfur Pollution and Research Approaches. Duke University, North Carolina, May, 1975.

4. _____ and Corn, M. "The Irritant Potency of Zinc Ammonium Sulfate of Different Particles Sizes." Industrial Hygiene Journal (1963), 326–33.

5. Atomic Energy Research Corporation. An Economic and Technical Appraisal of Air Pollution in the United Kingdom. Scientific Administrative Office, England.

6. Berman, M. "The Impact of Sulfur Oxide Pollution on Chronic Respiratory Disease." Unpublished paper, Yale University, 1976.

7. Binder, R. and Feinberg, D. "Estimates of Measurable Human Responses of Carbon Monoxide." ISPS Working Paper W4-7, Yale University, 1974.

8. Booz Allen and Hamilton, Inc. Study to Determine Residential Soiling Costs of Particulate Air Pollution. National Air Pollution Control Administration, Raleigh, North Carolina, 1970.

9. Brysson, R.J.; Trask, B.; Urban, J.; and Bouras, S. "The Effects of Air Pollution on Exposed Cotton Fabrics." Journal of Air Pollution Control Association, 17 (1967), 294.

10. Buechley, R.W., et al. "SO$_2$ Levels and Perturbations in Mortality: A Study in the New York-New Jersey Metropolis." Archives of Environmental Health, 27 (1973), 134–135.

11. _____, et al. "SO$_2$ Levels, 1967-72 and Perturbations in Mortality: A Further Study in the New York-New Jersey Metropolis." National Institute of Environmental Health Sciences, 1975.

12. Campbell, W.E. "Reduction of the Rate of Formation on Silver and Brass by Purification of the Atmosphere." Illinois Institute of Technology International Conference, 1972.

13. Charlson; Alquist, N.; and Horvath, H. "On the Generality of Correlation of Atmospheric Aerosol Mass Concentration and Light Scatter." _Atmospheric Environment_, 2 (1968), 455-464.

14. Costomis, A.C. and Sinclair, W.A. "Ozone Injury to Pinus Strobus." _Journal of Air Pollution Control Association_, 19 (1969), 867-71.

15. EPRI. "Sulfur Oxides: Current Status of Knowledge." Prepared by Greenway, Attaway, and Tyler, Inc., EPRI EA-450, May, 1977.

16. EPRI. "Evaluation of CHESS: New York Asthma Data 1970-71." Prepared by Greenway, Attaway, and Tyler, Inc., EPRI EA-316, December, 1976.

17. Ettinger, H.J. and Royer, G.W. "Visibility and Mass Concentration in a Non-Urban Environment." _Journal of Air Pollution Control Association_, 22 (1972), 108-111.

18. Gauri, K.L. and Sarma, A.C. "Controlled Weathering of Marble in a Dynamic SO_2 Atmosphere." _Proceedings of the 3rd Annual Meeting_. Louisville, Ky.: Environmental Engineering Science Conference, 1973.

19. Glasser, M. and Greenburg, L. "Air Pollution, Mortality, and Weather." _Archives of Environmental Health_, 22 (1971), 334.

20. Hammer, D.I., et al. "Los Angeles Student Nurse Study." _Archives of Environmental Health_, 28 (1974, 255-260.

21. Haynie, F.H. "Air Pollution Effects on Stress Induced Intergranular Corrosion of 7005-T 53 Aluminum Alloy." Draft, U.S. EPA, 1975.

22. _____ and Upham, J.B. "Effects of Atmospheric Sulfur Dioxide on the Corrosion of Zinc." _Material Prot. Perform._, 9 (1970), 35-40.

23. _____. _Effects of Atmospheric Pollutants on the Corrosion Behavior of Steel_. U.S. EPA, December, 1971.

24. Hershaft, A.; Morton, J.; and Shea, G. _Critical Review of Air Pollution Dose-Effect Functions_. For Council of Environmental Quality and U.S. EPA, 1976.

25. Hextor, A.C. and Goldsmith, J.R. "Carbon Monoxide: Association of Community Air Pollution with Mortality." _Science_, 172 (1971), 265.

26. Hodgson, J.A. "Short Term Effects of Air Pollution on Mortality in New York City." _Environmental Science and Technology_, 4 (1970), 589.

27. Horvath, H. and Noll, K. "The Relationship Between Atmospheric Light Scattering Coefficients and Visibility." Atmospheric Environment, 3 (1969), 543-552.

28. Kosmider, S. and Misiewicz, A. "Experimental and Epidemiological Investigations on the Effects of Nitrogen Oxides on Lipid Metabolism." Int. Arch. Arbeitsmed, 31 (1973), 249-56.

29. Lave, L.B. and Seskin, E.P. "An Analysis of the Association Between United States Mortality and Air Pollution." JASA, 68 (1973), 284-290.

30. _____. "Acute Relationships Among Daily Mortality, Air Pollution, and Climate." Economic Analysis of Environmental Problems. Edited by E. Mills. 1975.

31. _____. Air Pollution and Human Health. Baltimore: John Hopkins Press, 1978.

32. Logan,W. "Mortality from Fog in London, January 1956." British Medical Journal, I (1956), 722.

33. Mendelsohn, R. "Empirical Evidence on Home Improvements." Journal of Urban Economics, 4 (1977), 459-468.

34. _____ and G. Orcutt. "An Empirical Analysis of Air Pollution Dose Response Curves." Journal of Environmental and Management,

35. Michelson, I. and Tourin, B. Report on Study of Validity of Extension of Economic Effects of Air Pollution Damage from Upper Ohio River Valley to Washington, D.C. Area. Environmental Health and Safety Research Corp., 1967.

36. Miller, P.L. "Oxidant-Induced Community Change in a Mixed Conifer Forest." Air Pollution Damage to Vegetation. American Chemical Co., Washington, D.C., 1973.

37. Morrow, P.E. "An Evaluation of Recent NO_x Toxicity Data and an Attempt to Derive an Ambient Air Standard for NO_x by Established Toxicological Procedures." Environmental Research, 10 (1975), 92-112.

38. National Academy of Sciences. Air Quality and Stationary Source Emission Control. Prepared for the Committee on Public Works, U.S. Senate, 1974.

39. National Academy of Sciences. Air Quality and Automobile Emission Control. Prepared for the Committee on Public Works, U.S. Senate, 1974.

40. Ridker, R.G. Economic Costs of Air Pollution. New York: Frederick Prager, 1967.

41. Rylander, R. "Alteration of Lung Defense Mechanism Against Airborne Bacteria." Archives of Environmental Health, 18 (1969), 551-555.

42. Salvin, V.S., et al. "Advances in Theoretical and Practical Studies of Gas Fading." American Dyest. Reports, 14 (1952), 297-304.

43. Sereda, P.J. "Atmospheric Factors Affecting the Corrosion of Steel." Industrial and Engineering Chemistry, 52 (1960), 157.

44. Shy. "The Chattanooga School Children Study, Effects of Community Exposure to Nitrogen Dioxide, Part I and Part II." Journal of Air Pollution Control Association, 20 (1970), 539, 582.

45. Spence, J.F., et al. The Effects of Gaseous Pollutants on Paints: A Chamber Study. U.S. EPA, 1974.

46. Sprey, P. and Takacs. Health Effects of Air Pollutants and Their Interrelationship. U.S. EPA, 1974.

47. U.S. Environmental Protection Agency. Air Quality Criteria for Sulfur Oxides. Washington, D.C., 1969.

48. _____. Air Quality Criteria for Particulates. National Air Pollution Control Administration, AP-49, Washington, D.C., 1969.

49. _____. Air Quality Criteria for Nitrogen Oxides. National Air Pollution Control Administration, AP-84, Washington, D.C., 1969.

50. _____. Health Consequences of Sulfur Oxides: A Report From CHESS, 1970-71. National Environmental Research Center, Triangle Park, North Carolina, May, 1974.

51. Upham, J.B., et al. Fading of Selected Drapery Fabrics by Air Pollutants. U.S. EPA, 1974.

52. Waddell, T.E. The Economic Damages of Air Pollution. U.S. EPA, EPA-600/5-74-002, Washington, D.C., 1974.

53. Warner and Stevens. "Reevaluation of the Chattanooga School Children Study in Light of Other Contemporary Government Studies." Journal of Air Pollution Control Association, (1973), 769-772.

54. Winkelstein, W., et al. "The Relationship of Air Pollution and
       Economic Status to Total Mortality and Selected Respiratory
       System Mortality in Men, Part I and Part II: Suspended
       Particulates and Sulfur Oxides," Archives of Environmental
       Health, 16 (1968), 401; 20 (1971), 1974.

55. Zeidberg. "The Nashville Air Pollution Study: Mortality from
       Diseases of The Repiratory System in Relation to Air Pollution."
       Archives of Environmental Health, 15 (1967), 214; 15 (1967),
       255.

CHAPTER VI

INTRODUCTION TO THE SIMULATIONS

In order to illustrate the usefulness of the environmental model
discussed in Chapters II through V, the model is applied to an analysis
of air pollution generated by a hypothetical source. The source is a
coal-fired 500 MW electrical generating station near New Haven,
Connecticut. The environmental model is calibrated with data describing
this source and location. The consequences of several air pollution
abatement techniques are simulated with the calibrated model. These
consequences are reviewed and compared in Chapters VII and VIII.

A number of assumptions are made in the process of applying the
general environmental model to this particular case study. Each of the
steps of the model shown in Figure II-A are adapted to the data for the
case study. The necessary assumptions behind this adaptation are
identified and discussed in the remainder of this chapter.

Because the hypothetical plant is supposed to be new, the station is
assumed to be able to accomodate each air pollution abatement technique
without special installation costs. The problem of retrofitting or
relocating existing plants is not explored in this analysis. The plant
is assumed to generate electricity 6500 hours a year (about 75% of the
time). Because shut downs for repairs are random with respect to weather
conditions, it is assumed that the electricity is generated at a constant
rate for the entire year. Though plant operation may vary systematically
with the time of day (more kilowatt hours during peak periods), it is
further assumed that this plant operates at a constant rate all day. The
effect of peak period operations on air pollution control is an important

topic but it is beyond the scope of this study.

The load center for the electricity generated by the power station is assumed to be downtown New Haven. Electricity generated outside of New Haven must, therefore, be shipped to the downtown area. The extra cost of locating a plant out of the downtown area is assumed to be the costs of transporting this electricity back to New Haven on overhead transmission lines. Because of aesthetic objections and possible health hazards, some citizens may prefer the use of underground transmission lines. This would significantly raise the cost of remote siting (see Chapter III for more details). This study ignores spatial variations in the price of the land necessary for the power plant. Because land is often cheaper in more distant locations than in New Haven harbor, this potentially overstates the cost of remote siting. The cost of the land for the power station, however, is a small part of the total cost of generating electricity and the variations in land prices across southern Connecticut are not severe enough to make a significant difference.

The cost and effectiveness of each of the abatement techniques discussed in Chapter III may vary in different regions. For instance, arid regions, because of higher evaporation rates, permit the use of slightly different flue gas desulfurization systems than New Haven does. It is assumed that the expected costs and performances of the abatement systems described in Chapter III apply to the New Haven region. These figures are presented in Tables III-2,3,4,5,7,11, and 14.

The meteorological model described in Chapter IV is calibrated with data from Bradley Field in Hartford, Connecticut. This data describes the annual frequency of winds from each direction as well as the wind speeds, cloud cover, and ceiling height. It is assumed in the calculations of dispersion that the climatic variables which are measured in Hartford apply throughout New England. Although this assumption is not realistic for locations which are a considerable distance from Hartford, these errors are not expected to have significant ramifications upon the analysis. The inland measurements of Hartford weather, however, fail to capture the effects the Long Island Sound may have upon a coastal area such as New Haven. The effects of the small hills around New Haven are also ignored. The omission of these topological influences decreases the accuracy of the model's predictions. On the other hand, ignoring these effects should not result in any systematic biases.

In order to calculate general population exposures, it is necessary to estimate the number of people breathing pollutants at each location. Because pollutant concentrations are different inside and outside buildings, it is difficult to calculate exposures precisely. For the purposes of the simulations, people are assumed to breathe the ambient ground level concentrations of pollutants near their place of residence. Of course, many people spend a considerable amount of time outside their residence, especially working. The total exposures in areas with higher ratios of jobs to residents is, therefore, underestimated by the model. The population exposures from pollution in downtown areas, such as in New Haven, may be underestimated slightly. This factor provides a small bias against the value of remote siting and higher effective emission heights. On the other hand, exposures to persons in outdoor recreational areas are probably underestimated by the model. The net effect of both sources of bias are probably unimportant.

The population at each location is calculated from 1970 U.S. Census estimates. The number of people in each sex and in three broad age categories is calculated for each area. The number of people in three income groups, two races, and relevant states are also computed for each town in order to calculate equity effects. Exposures near the power plant site are computed on a town by town basis. For local areas, it is assumed that everyone in the town is exposed to the average level of pollution of that town. In order to reduce the cost of running the model, more distant towns are aggregated into larger spatial units. For example, all the residents of Delaware are assumed to be exposed to the average pollution concentration in Delaware. Because the pollution gradient is not very steep beyond 80 kilometers, the approximations from these aggregations are not expected to seriously effect the accuracy of the results.

The dose response functions are described in detail in Chapter V. The mortality figures for each age and sex group in Table V-3 are used to calculate premature deaths. The morbidity figures in Tables V-6 and V-7 are used to calculate chronic and acute illnesses, respectively. Because there are no spatial estimates of damageable materials, it is assumed that these materials are distributed in proportion to the local population. The material dose response estimates in Table V-1 are used to convert material exposures into damages. Damages from visibility are also assumed to be proportional to the population in each area. The amount of visibility loss is calculated from figures presented in Chapter V. The dollar value of these losses is discussed in greater detail in Chapter VII. The damages from acid rain and vegetation losses are assumed to be proportional to the level of emissions. Both vegetation and acid rain damages are calculated directly from emissions. The vegetation damage functions are presented in Table V-2. Acid rain effects are discussed in the text of Chapter V.

CHAPTER VII

A COMPARISON OF THE EFFECTIVENESS OF ABATEMENT TECHNIQUES

One of the most important parameters which distinguish abatement techniques from one another is effectiveness. What will a dollar spent on a particular abatement technique do to aggregate air pollution damages? Through a series of simulations upon the environmental-economic model described in Chapter II, the effectiveness of several abatement technologies are estimated. These estimates of effectiveness are compared and analyzed in the remainder of this chapter.

In order to understand the benefits of pollution control measures, it is helpful to understand the nature of the air pollution damages they are designed to eliminate. Table VII-1 presents a list of the expected damages from the uncontrolled emissions of a 500 MW power plant located in New Haven harbor. The estimates in Table VII-1 include damages which occur in the entire northeastern United States, not just in New Haven itself. Total dollar damages to vegetation plus materials is expected to be 1.5 million dollars a year. Visibility is reduced by 3.4 million person-kilometer-years (an amount equivalent to 3.4 million persons losing one kilometer of visibility for an entire year). Annually, over 40 lives are expected to be terminated prematurely. Though all of these excess deaths represent a loss of expected lifetime, only one-third of these fatalities strike people under 65. The air pollution is also associated with over 1800 cases of bronchitis, pneumonia, croup, or lower respiratory disease. Each of these cases are expected to inconvenience a patient about 90 days on the average. Finally, there are about 20,000 cases of acute health effects which provide symptoms lasting about one

Table VII-1

Predicted Annual Damages of the Uncontrolled Emissions of a 500 MW Coal-Fired Power Plant in New Haven Harbor

Consequence	Total	Sulfur Dioxide	Particulates	Nitrogen Oxide	Carbon Monoxide	Sulfate
Visibility (10^6 people kilometer years)	3.42 (1.4)	.19 (.1)	3.18 (1.4)	.02 (.01)	--	.03 (.01)
Acid Rain (10^3 \$)	770 (361)	712 (360)	--	44 (23)	--	14 (8)
Materials (10^3 \$)	584 (522)	459 (438)	101 (276)	15 (55)	--	9 (6)
Vegetation (10^3 \$)	197 (184)	149 (180)	.00	44 (45)	--	3 (3)
Deaths:						
18-44 year olds	3.3 (1.3)	2.4 (1)	.6 (.4)	.1 (.5)	.00 (.0)	.3 (.2)
45-64 year olds	9.1 (6)	14.1 (4)	-5.6 (5)	-.7 (-.5)	.00 (.0)	1.2 (.4)
65+ year olds	30.7 (13)	26.9 (10)	2.6 (9)	-2.0 (1)	.00 (.0)	3.3 (1)
Three Month Illnesses	1,856 (1,500)	967 (1,300)	764 (795)	--	--	127 (85)
Acute Illnesses	23,620 (42,000)	8,640 (8,400)	13,400 (41,000)	690 (300)	6 (4)	830 (800)
Quantity Emitted (10^3 tons)	250	71.2 (10)	165.8 (30)	11.1 (3)	.6 (.1)	1.4 (.6)

day. These include headache, eye irritation, chest discomfort, and
asthma attacks. The sum of the expected active days lost because of
this one 500 MW coal boiler is over 400,000 days annually. If loss of
visibility were valued at $.50 per person-kilometer-year, total non-
health damages would be worth over three million dollars a year. If a
day lost were valued at $25 a day, total health damages would amount to
about 10 million dollars annually. Though the true sizes of these
aggregate dollar amounts depend upon the evaluation of health and
visibility losses, given reasonable values for these losses, air
pollution damage is far from negligible.

In addition to presenting expected losses, Table VII-1 also presents
a measure of the degree of certainty of each predicted value. The values
in parentheses correspond to standard errors for each expected value.
Though the standard deviations presented here do not capture all the
implicit uncertainty in the model (a near impossible task!), they provide
information about the reliability of each estimate. The ratio of the
expected value to its standard error, for instance, is one commonly used
measure of a parameter's reliability.[1] The standard errors of the
estimates of material and vegetation damages are almost as large as the
estimates themselves. The real effects may well be near zero or much
larger than the expected estimate. On the other hand, the estimates of
visibility loss and mortality (except for 45-64 year olds) appear to be
relatively precise. Rough estimates of the variance of the aggregate
figures suggest that both the total health and non-health damages of a
coal fired power plant are statistically different from zero and less
than twice the expected estimate (the ratio of these expected values to
their standard errors is about three for both figures).

Another important facet of air pollution abatement control is the allocation of total damages to specific types of emissions. Two-thirds by weight of all emissions are particulates (not including sulfate). They account for one-half of non-health damages but only 11% of health effects. Sulfur dioxide emissions account for only 28% of emissions by weight but 43% of non-health damages and over four-fifths of days lost. The sulfur dioxide emissions probably cause more harm than all the other emissions combined from an uncontrolled coal-fired boiler. Not all of these damages are caused by ambient sulfur dioxide—a large proportion of the health effects are actually caused by sulfate, a secondary product of the sulfur dioxide emissions. Carbon monoxide, though toxic, contributes negligibly to total damages because large boilers emit only small quantities of this gas. Damages from nitrogen oxide emissions are limited primarily to vegetation. Ozone formation downwind of the power plant site is responsible for most of these damages. Because sulfates are not emitted in great quantities, their aggregate impact is relatively small: 1% of total non-health damages and 9% of all days lost. Sulfate emissions, however, should not be ignored because only small increases in sulfate emissions can result in extensive damage.

The choice of abatement technique not only effects the level of damages, it also effects the type of damages still caused by the power plant. Nitrogen oxide control primarily decreases damages connected with acid rain and vegetation. Carbon monoxide control will reduce health effects. The reduction of particulates primarily alters visibility and mortality rates. Only reducing sulfur emissions has an across-the-board impact upon all types of air pollution damage. Even sulfur control, though, has sharper impacts on some types of damage than others.

Abatement techniques which prevent emissions will lower final damages in direct proportion to the reduction of each pollutant. For example, a ten percent reduction of sulfur dioxide emissions will reduce the air pollution damages caused by sulfur dioxide by ten percent. The benefits of dispersion oriented abatement techniques involve a more complicated estimation than the benefits of emission oriented control techniques. Figure VII-A demonstrates the total health losses that a 500 MW coal-fired power plant will cause in several locations around Connecticut. Locations close to major cities result in significantly higher exposures and final effects as well. Because of the larger populations towards the western end of the state and because of the prevailing westerly winds, the level of health effects generally falls as one moves an emission source eastwards. The best direction to move a plant out of New Haven is east-south-east. This direction quickly locates the plant in or across Long Island Sound. Because of the difficulty of locating a plant on water or of transporting electricity across such a large body of water, the east-south-east direction is really not practical. The next best direction out of New Haven is due east. All the sites in the following analysis are either in New Haven or due east because there is no reason to choose an inferior direction. The optimal distance of a power plant from New Haven is a complex problem. The total damage of a given level of emissions clearly falls with distance east of New Haven. Unlike emission controls, however, the benefits of additional distance have a diminishing impact upon damages as distance increases. The first ten kilomters east of New Haven are as effective in reducing total damages as the next fifty kilometers. Using higher stacks, a stationary dispersion technique, also has diminishing returns with additional height.

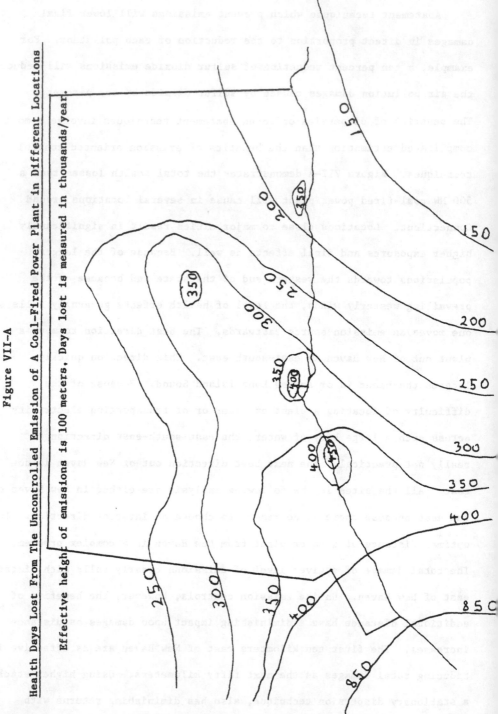

Figure VII-A

Health Days Lost From The Uncontrolled Emission of A Coal-Fired Power Plant in Different Locations

Effective height of emissions is 100 meters, Days lost is measured in thousands/year.

Evaluating all of these abatement techniques involves comparisons across different bundles of goods. In order, for example, to compare one particulate control device against another nitrogen oxide control device, one must inevitably weigh improvements in visibility against vegetation and acid rain damage. In order to evaluate the abatement techniques, it is necessary to develop an index or common unit of measurement by which to compare dissimilar objects or effects. The problem of developing appropriate indices for bundles of goods is a familiar dilemna in economics.[2] Microeconomics tells us that for a bundle of known goods bought in perfectly competitive markets, the "best" index of these goods is a weighting using their market prices. However, most of the benefits from air pollution control are simply not exchanged in any market. One cannot buy deaths, morbidity, or visibility in any legitimate market. In an effort to evaluate untraded goods, economists have developed methodologies to determine pseudo or shadow prices for goods. Shadow prices are constructed from the cost of a traded good or a set of traded goods which produce comparable services to the untraded goods. For example, the shadow price for the value of life could be determined from how much people demand in premium wages in order to perform risky jobs. The shadow price for visibility could be determined from the relative wage levels in cities with clear versus polluted skies. Shadow prices, however, are not as reliable as known market prices since they make an analogy amongst goods which may or may not be appropriate. In addition, frequently there are serious estimation problems with shadow prices.

A third difficulty with the use of market prices as an index for pollution benefits is that they must be shared at a common level. One of the primary reasons why market prices are considered a good index is because they reflect the choices of every individual buying what he wants to own. Air pollution benefits, however, are distributed across individuals without regard to their tastes or income.[3] Air pollution benefits, therefore, should be evaluated in terms of their value to the average individual, not the individual on the margin. Because the marginal buyer is always willing to spend more for the marginal good than the average individual, market prices overestimate the value of these commonly shared benefits. The market price of death is an overestimate of the average values. The value of life to the average individual is greater than the price indicated by wages for risky jobs.

The thrust of the above discussion is that one cannot aggregate the various effects of air pollution into a single index without making strong assumptions or value judgments. A model completely free of such judgments would have to report all the effects of the power plant's emissions in terms of their original physical units. Even the categories in Table VII-1 are aggregates of more micro units so that the final list would actually have to include thousands of entries. This list could easily be so overwhelming, however, that the important patterns across all the entries would be lost from sight. In order to make the results more accessible to the reader, the myriad effects of power plants upon man, society, and nature are aggregated into two broad categories: health and non-health effects. These two categories are not combined into a final aggregate damage level because the value of health is too important a value judgment to be made by a scientist, economist, or a model builder. It is the value of a day of health which is at the crux of the evaluation of air pollution damages. The remaining value judgments which are necessary to arrive at health and non-health totals are relatively less controversial and they also have a smaller impact on the final evaluation.

The dollar estimates of total vegetation, material, and acid rain losses are calculated by valuing goods at their market prices. Because most of these damaged goods are traded in the market, the amount of information lost by aggregating these dollar estimates into one single dollar figure is not expected to be large. Estimating the market value of visibility losses, on the other hand, cannot be done as precisely. The best information one has about the value of visibility comes from survey data. There is little guarantee that survey results reflect the true values of individuals. As can be seen in Table VII-2, the evaluation of visibility losses can have a major impact on the size of non-health losses. The analysis in the following pages assumes that visibility is valued at $.50 per lost kilometer per year by every affected individual.

In order to aggregate very different health effects, the value of each health effect is assumed to be equal to the expected active days lost per case of each effect. A bad headache which reduces activity for one day is thus valued at one. All the acute effects except premature death are valued at one day lost per case. As discussed in Chapter V, deaths due to acute as opposed to chronic exposures are expected to be premature by no more than one year. Each of these deaths is assumed to cause the loss of about 90 days. Bronchitis, lower respiratory disease, pneumonia, and croup are also expected to incapacitate a patient an average 90 days per case. The active days lost associated with chronic deaths are assumed to be equal to the expected lifetime of each

Table VII-2

The Sensitivity of Total Non-Health Damages
to the Price of Visibility[a]

	Price of Visibility $ per person-kilometer					
	0	.01	.25	.50	1.0	5.0
Total Non-Health Damages (10^3\$)	1551	1585	2406	3261	4971	18651

[a] These totals are for a single 500 MW coal-fired power plant in
New Haven harbor.

individual.[4] For persons 18-44 years old, this loss is equal to 40

years. Middle aged adults lose about 25 years and people over 65 are

assumed to lose about 10 years of active days if prematurely killed.

Because the relative value of each health effect is assumed equal to the

resultant active days lost, the health effects can be aggregated in terms

of total days lost. The dollar value of total health effects can,

therefore, be discussed in terms of what one day of good health is

worth. The dollar value of each health effect is simply the dollar value

of an active day times the active days lost. For example, at $25 per

day, a headache is worth $25 and the premature death of a person 35 years

old is worth $365,000.

Comparing technologies on the basis of their expected health and

non-health costs, it is apparent that some techniques are better than

others. That is, for every level of health benefits, there exists a

technology or set of technologies which achieves that level of benefits

at the lowest level of non-health costs. These technologies are the most

effective. They make up the production possibility frontier for air

pollution control. Table VII-3 contains a list of abatement technologies

and their expected effects. The techniques not underlined are less

effective because they incur unnecessarily higher costs in order to

achieve a given reduction in health effects. Using just low sulfur fuel,

for example, is an inferior technique because using beneficiated (to

Level 3) coal and a high stack (an additional 40 meters) reduces health

losses by over 40,000 additional days and costs three million dollars

less annually.

Figure VII-B plots the expected health benefits versus the expected

non-health costs (the abatement costs plus the non-health damages) of a

Table VII-3

Partial List of Abatement Techniques and Their Annual Aggregate Effects[a]

(underlined techniques are effective)

Technique	Active Days Lost (days x 10³)	Abatement Cost ($ x 10⁶)	Non-Material Damages Plus Abatement Cost ($ x 10⁶)
Nitrogen Oxide Flue Gas Removal in New Haven	444.9 (159)	7.15 (2.0)	10.31 (2.2)
Staged Combustion in New Haven	437.8 (159)	0 (1.0)	3.22 (1.4)
No Abatement in New Haven	433.3 (159)	0 (0)	3.26 (1.0)
Particulate Cyclone in New Haven	413.9 (140)	6.5 (1.0)	9.08 (1.3)
Electrostatic Precipitator 97% Removal in New Haven	385.9 (130)	4.55 (1.0)	6.29 (1.2)
Electrostatic Precipitator 99% Removal in New Haven	383.4 (130)	12.0 (2.0)	13.68 (2.1)
Venturi Scrubber (High) in New Haven	379.4 (130)	12.0 (1.5)	13.78 (1.7)
Air Bag Filter (A/C = 2) in New Haven	370.0 (126)	9.4 (2.0)	11.01 (2.1)
High Stack (+20 meters) in New Haven	352.3 (140)	.01 (.01)	2.39 (.7)

(cont.)

(cont.)

Technique	Active Days Lost (days x 10³)	Abatement Cost ($ x 10⁶)	Non-Material Damages Plus Abatement Cost ($ x 10⁶)
High Stack (+40 meters) in New Haven	310.3 (130)	.03 (.02)	1.98 (.5)
Locate 10 km East of New Haven	272.7 (120)	.80 (.2)	2.31 (.5)
Electrostatic Precipitator and Locate 10 km East	263.3 (120)	5.35 (1.4)	6.56 (1.6)
Locate 15 km East	253.6 (110)	1.2 (.3)	2.61 (.5)
Air Bag Filter Plus Locate 15 km East	243.4 (110)	10.62 (2.0)	11.77 (2.1)
Coal Beneficiation-Level 3 Power Plant in New Haven	231.7 (80)	2.93 (.8)	4.48 (.99)
Locate 25 km East	226.2 (100)	2.0 (.5)	3.33 (.70)
Low Sulfur Fuel in New Haven	198.3 (130)	17.90 (2.0)	20.64 (2.2)
Meyer Chemical Coal Cleaning in New Haven	187.4 (110)	32.8 (8.0)	35.16 (8.1)
Beneficiation-Level 4 in New Haven	181.9 (70)	7.80 (1.8)	9.04 (1.8)
Locate 50 km East	181.6 (90)	4.00 (1.0)	5.24 (1.1)
Beneficiation-Level 3 Plus High Stack (+40 meters) in New Haven	170.3 (70)	2.96 (.8)	3.98 (.9)

(cont.)

(cont.)

Technique	Active Days Lost (days x 10³)	Abatement Cost ($ x 10⁶)	Non-Material Damages Plus Abatement Cost ($ x 10⁶)
Locate 75 km East	156.0 (80)	6.0 (1.5)	7.18 (1.6)
Beneficiation-Level 3 Plus 25 km East	126.5 (60)	4.93 (.99)	5.69 (.9)
Hydro Chemical Coal Cleaning Power Plant In New Haven	108.9 (100)	39.6 (9.0)	41.07 (9.0)
Flue Gas Desulfurization in New Haven	105.6 (90)	15.90 (2.0)	17.74 (2.1)
Beneficiation-Level 3 Plus 75 km East	87.5 (50)	6.93 (1.7)	7.65 (1.72)
Low Sulfur Fuel Plus 50 km East	71.0 (40)	21.9 (2.2)	22.6 (2.2)
Flue Gas Desulfurization Plus Electrostatic Precipitator 97% in New Haven	61.5 (20)	20.45 (2.2)	20.81 (2.2)
Beneficiation-Level 3 Plus Flue Gas Desulfurization in New Haven	31.2 (10)	18.83 (2.25)	19.08 (2.2)
Beneficiation-Level 3 Plus Flue Gas Desulfurization Plus High Stacks (+40 meters) in New Haven	23.5 (20)	18.86 (2.2)	19.08 (2.2)

(cont.)

(cont.)

Technique	Active Days Lost (days x 10³)	Abatement Cost ($ x 10⁶)	Non-Material Damges Plus Abatement Cost ($ x 10⁶)
Beneficiation—Level 3 Plus Flue Gas Desulfurization Located 25 km East	18.0 (20)	20.83 (2.2)	21.08 (2.2)
Beneficiation—Level 3 Plus Flue Gas Desulfurization Located 50 km East	14.4 (20)	22.83 (2.4)	22.99 (2.4)
Low Sulfur Fuel Plus Flue Gas Desulfurization Plus Air Bag (A/C = 2) in New Haven	13.1 (10)	43.22 (3.5)	43.40 (3.5)
Beneficiation—Level 4 Plus Flue Gas Desulfurization Located 75 km East	9.6 (20)	27.70 (3.1)	27.87 (3.1)
Low Sulfur Fuel Plus Air Bag Filter (A/C = 2) Plus Flue Gas Desulfurization Located 75 km East	7.3 (20)	49.22 (3.8)	49.37 (3.8)

[a] Nitrogen oxides are predicted to have positive effects on health in Chapter V. Though this result is quite uncertain, it results in the poor rating of nitrogen oxide control techniques.

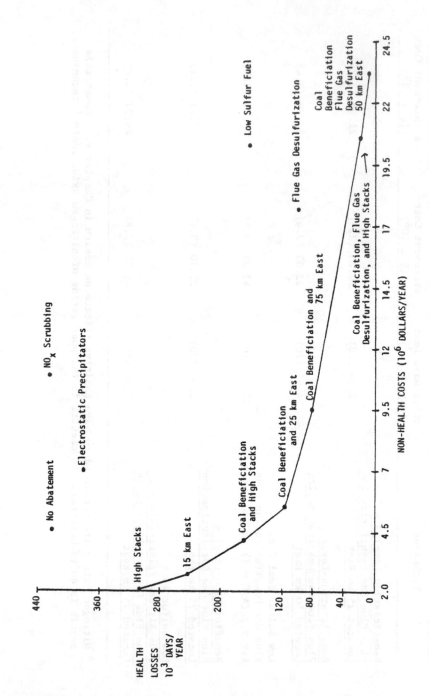

FIGURE VII-B

THE TRADEOFFS BETWEEN HEALTH AND NON-HEALTH COSTS

broad spectrum of abatement techniques. The solid line sloping down
from the left represents the locus of efficient technologies. All
points to the right of this curve are less effective because they entail
unnecessary non-health costs for the level of health benefits that they
provide. The horizontal distance from these points to the curve is equal
to the amount of money wasted by these technologies to achieve a given
level of health effects. The vertical distance from these points to the
curve is the expected number of days lost which could have been averted
if the same non-health costs had been allocated to the most efficient
technique.

All of the most efficient techniques utilize some form of dispersion.
Every one of the most effective abatement processes either utilizes
higher stacks or locates the power plants outside of the central city.
Dispersion alone is the most effective available technique for levels of
health effects of 200,000 or more days per year. For levels of health
effects below 200,000 days per year, it is more effective to use coal
beneficiation and eventually flue gas desulfurization along with dispersion.
Pure hardware solutions, in contrast, are inefficient. Particulate
removal systems cost between 5 and 10 million dollars more per year than
higher stacks while higher stacks save over 60,000 more health days per
year. Nitrogen oxide removal systems are not expected to reduce health
losses at all but yet they may cost between 2 and 8 million dollars more
than higher stacks. Flue gas desulfurization systems may cost between
10 and 20 million dollars more per year than a combination of dispersion
and coal beneficiation but yet achieve the same health reductions.
Dispersion is an important component of effective air pollution control
measures.

As can be seen in Table VII-4, the marginal cost of additional
health reductions rises rapidly as one attempts to eliminate health
effects. Diminishing returns is especially evident within a given type
of abatement. Locating the power plant east of New Haven is very
effective at first, but as the plant gets further away, this effectiveness
clearly diminishes. Additional expenditures on hardware systems are also
less effective than the initial expenditures. For instance,
electrostatic precipitators which remove 99% of particulates are twice
as expensive as those which only remove 97%, but the more expensive
system only lowers health effects by an additional 5%.

The marginal effectiveness of most single technologies also
diminish when combined with other abatement techniques. Coal
beneficiation (Level 3), for example, reduces expected health losses by
200,000 days if used alone. In conjunction with locating the plant
25 km east of the central city, the marginal reduction due to coal
beneficiation is 100,000 days. In conjunction with locating the plant
25 km east of the central city and using flue gas desulfurization, the
marginal impact of coal beneficiation is just 25,000 days. Adding
hardware devices to eliminate emissions reduces the effectiveness of
locating a plant away from population centers. Uniform regulations
designed for plants in central cities will not only require too much
hardware at remote sites, but they will also remove the incentive to site
plants effectively in remote areas.

Identifying the set of technologies which minimize non-health costs
for every level of health effects is largely a technical question.
Choosing the best technology from this set is largely a value judgment.
In order to make the best judgment, however, it is helpful to know more

about the available choices. One important difference among these
remaining technologies is the amount of money being spent to eliminate a
health effect. The marginal cost of reducing health effects is
presented in Table VII-4. As mentioned earlier, the marginal cost of
reducing health effects rises rapidly as total health effects are
reduced. If society chooses to have the lowest possible level of health
effects, the cost of the last few active days lost will rise as high as
8400 dollars per day. At this cost per day, the life of a 35 year old
would be worth 120 million dollars. The cost of removing as many of the
health effects as possible might be more than society is willing to pay.
How many active days should be lost depends upon what society is willing
to pay to avoid the loss of a marginal active day.[5]

Each of the effective technologies is most efficient for particular
values of a day lost. For low values of life, the most efficient
technique to control air pollution from a coal-fired power plant for New
Haven is to locate it 15 km east of New Haven. For values above $16 a
day, the most efficient techniques involve a combination of coal
beneficiation at Level 3 and high stacks or remote siting. Unless the
value of a day exceeds $150, it is not efficient to add any hardware to
the power plant at all. Flue gas desulfurization in combination with
other techniques becomes efficient only with values above $150 a day.
Particulate removal systems do not become efficient until the value of
life exceeds $2000 a day. These results, it should be kept in mind, are
specific to New Haven, but they at least raise questions about the
adoption of pure hardware solutions to the problem of air pollution.

It is not clear that one would want to judge the effectiveness of a
technique solely with its expected values. The   ovals in Figure VII-C

Table VII-4

The Marginal Cost of the Last Day Lost Using Effective Techniques

Treatment	Total Days Lost (10^3 days)	Abatement Costs Plus Non-Health Damages (10^6 $)	Marginal Cost Per Day ($/day)
High Stacks (40 m)	310.3	1.98	-
Remote Siting: 10 km East	272.7	2.31	8.78
15 km East	253.6	2.61	15.71
Ben 3 + High Stacks (40 m)	170.3	3.98	16.45
Ben 3 + Remote Siting: 25 km East	126.5	5.69	39.04
50 km East	101.7	7.65	79.03
75 km East	87.5	8.93	90.14
Ben 3 + FGD + High Stacks (40 m)	23.5	19.08	158.59
Ben 3 + FGD + Remote Siting:			
25 km East	18.0	21.03	354.54
50 km East	14.4	22.99	544.44
75 km East	9.6	29.87	1433.33
Low Sulfur + Air Bag Filter + FGD + 75 km East	7.3	49.37	8478.26

Ben 3 = Beneficiation-Level 3   FGD = Flue Gas Desulfurization

[a] The total days lost with no abatement is 433,300 days per year and the total non-health damages are $3,260,000 per year. The total non-health costs of higher stacks are less than with no abatement because the abatement costs of higher stacks are not as great as the reductions in non-health damages.

Figure VII-C

The Uncertainty of the Estimated Effects[a]
of Selected Abatement Techniques

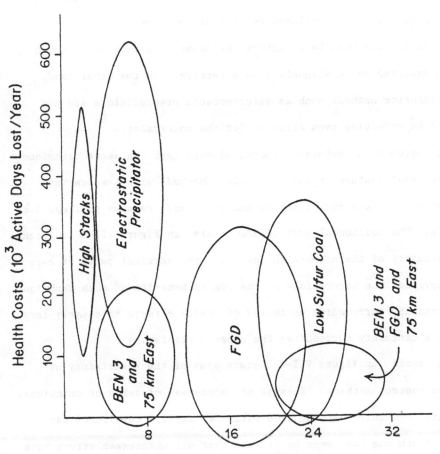

a The ovals measure two standard deviations around the expected
effects. BEN 3 is coal beneficiation (level 3) and FGD is
flue gas desulfurization.

which are two standard deviations in width, measure some of the
uncertainty surrounding the estimates of health and non-health costs.
Techniques which are close to the production possibility frontier may,
given the uncertainty, be just as effective as the best techniques. For
example, remote siting 25 kilometers east of New Haven or flue gas
desulfurization may both be effective abatement processes even though
they are expected to be slightly less effective. On the other hand,
clearly inferior methods such as electrostatic precipitators are not
likely to be effective even allowing for the uncertainty.

The degree of uncertainty associated with each abatement technique
is an important feature in and of itself. Methods with less certain
costs or benefits are more risky to undertake and, ceterus paribus, less
desirable. The horizontal part of the ovals in Figure VII-C represent
the uncertainty of the non-health costs and the vertical parts of each
oval represent the uncertainty of the health benefits of each technique.
The uncertainty surrounding estimates of health effects is clearly larger
than the uncertainty surrounding the other estimates.

The ovals in Figure VII-C capture some of the uncertainty of
pollution control methods. Because of unforeseen mistakes or omissions,
it is still possible that the true estimates may lie outside of each
oval. Though one can never be certain that all unforeseen errors have
been accounted for, sensitivity tests can provide a sense of the
robustness of the results. Two sensitivity tests are performed upon the
dose response functions because they are suspected to be the weakest link
in the model. The first sensitivity test uses only positive and
significant dose response functions. In other words, the first
sensitivity test explores the importance of the use of insignificant and

negative dose response functions.   The second test run uses Lave and
Seskin's [1] results instead of the results of Mendelsohn and Orcutt [2].
These new simulations depend upon a different body of data and
alternative analyses of health effects than the original simulation.

The marginal costs of health reductions for each of these
sensitivity runs are displayed in Tables VII-5 and VII-6.   The marginal
cost of health associated with each technology is virtually the same
with the sensitivity runs and the original simulation.With respect to
their marginal cost, the relative order of the technologies hardly
changes.   Also, dispersion is a part of every effective combination of
abatement technologies in all runs.   The only noticeable difference
brought about by the use of these alternative dose response curves is the
level of the   marginal cost of eliminating health effects.   The results
of the environmental-economic model are robust with respect to at least
some different estimates of man's responses to pollutant exposures.

Table VII-5

The Marginal Cost of Health Reductions––
Sensitivity Test One – Only Significant Dose Response Curves[a]

	Active Days Lost (10^3 days)	Non-Health Cost (10^6 \$)	Marginal Cost Per Day ($/Lost Day Prevented)
Higher Stacks[b]	256.8	2.40	–
Ben 3 + Higher Stacks[b]	147.0	4.01	14.7
Ben 3 + Remote Siting: 25 km	113.3	5.80	53.1
75 km	78.8	7.74	56.2
Ben 3 + FGD + Higher Stacks[b]	26.8	19.13	219.0
Ben 3 + FGD + Remote Siting: 25 km	19.5	21.09	268.5
75 km	13.5	27.96	1145.0

Ben 3 = Beneficiation-Level 3   FGD = Flue Gas Desulfurization

[a] The only difference between these calculations and those for Table VII-4 is that only significant dose response curves are used in this table.

[b] Higher stacks refers to raising the stack height 40 meters.

Table VII-6

The Marginal Cost of Health Reductions—
Sensitivity Test Two – The Mortality Dose Response Curves of Lave and Seskin[a]

	Active Days Lost (10³ days)	Non-Health Cost ($ 10⁶)	Marginal Cost Per Day ($/Lost Day Prevented)
Higher Stacks[b]	211.8	1.98	–
Remote Siting: 15 km	125.1	2.61	7.3
25 km	107.3	3.33	40.4
Ben 3 + Remote Siting: 25 km	53.8	5.69	44.1
75 km	35.3	7.63	104.9
Ben 3 + FGD + Higher Stacks[b]	10.2	19.07	455.8
Ben 3 + FGD + Remote Siting: 25 km	7.4	21.02	696.4
Ben 4 + FGD + Remote Siting: 75 km	4.2	27.87	2140.0

Ben 3, 4 = Beneficiation-(Level 3,4), respectively   FGD = Flue Gas Desulfurization

a The only difference between this table and Table VII-4 is the use of mortality dose response curves from Lave and Seskin [1] instead of Mendelsohn and Orcutt [2]. These new coefficients are shown in Table V-4.

b Higher stacks refer to raising stack height 40 meters.

Footnotes

[1] If the probability distribution around these estimates is Gaussian, the ratio of the expected value to the standard error is equivalent to a t-test. In this case, if the ratio is above two, the expected estimate is statistically different from zero. Also, twice the standard error would be the width of the 95% confidence interval. Such confidence intervals should be interpreted with caution, though, because the probability distribution may not be Gaussian and the estimates of the standard errors are crude.

[2] The familiar index problem in economics is how to measure the gross national output. There appears to be no perfect measure for comparisons among countries or across time.

[3] Because air pollution tends to decline with distance from the city, it would appear that people can buy different levels of pollution depending on their taste and ability to pay. This choice, however, locks an individual into residing in restricted locations. Further, changes in air quality can be evaluated at the margin only if migration is costless or if these changes are perfectly anticipated. If cities could be rearranged at no cost, the benefits could be evaluated in terms of what the marginal individual pays.

[4] The expected days lost from illnesses are reviewed in NAS [3], p. 611. The expected days lost from deaths are assumed to be equal to the average lifetime of an individual who is already a certain age.

[5] There are many decisions across society which entail determining how much a life is worth. Though these other decisions may be relevant to the question at hand, there may be important differences between a life lost from air pollution versus some other cause. For instance, air pollution exposures are largely involuntary. Health effects from air pollution may seem worse than similar effects caused by cigarette smoking. The timing of the health loss may also be important. Immediate illnesses may be valued differently from latent illnesses. The number of people affected at one time may also play a role. Large calamities seem to attract more attention than scattered deaths even though the same number of people die. Whether the victim is identifiable or anonymous before the decision and afterwards may also be important. Society may be loathe to kill handpicked individuals in order to secure some hedonistic pleasure. It may also be more sympathetic with the pleas of individuals who are clearly victims of a decision than with a statistical increase in health losses. Because the circumstances surrounding a death may be important, how much should be spent on air pollution deaths need not be equal to the money spent to avoid deaths from other causes.

[6] The health effects from air pollution are probably borne largely by the victim because it is never clear that air pollution is the cause. The victim may share medical costs but he cannot obtain full compensation without proving adequate cause.

Bibliography

1. Lave, L. and Seskin, E. Air Pollution and Human Health. Baltimore: Johns Hopkins Press, 1978.

2. Mendelsohn, R. and Orcutt, G. "An Empirical Analysis of Air Pollution Dose Response Curves." Journal of Environmental Economics and Management, (1979).

3. National Academy of Sciences. Air Quality and Stationary Source Control. Prepared for Public Works Committee of the U.S. Senate, Washington, D.C., 1975.

CHAPTER VIII

CONSIDERATIONS OF EQUITY

Though the effectiveness of various abatement technologies is an
important parameter for the management of our scarce environmental
resources, the distribution of these costs and services across
individuals is also important. The distribution of all goods and
services across income groups is one dimension of equity which has
received substantial attention. Another dimension of equity which has
often been studied involves the racial black/white bias of a particular
program or policy.[1] A third level of equity which is peculiar to
externality problems is whether the producer of the pollution, the
consumer of the good which the polluter produces, or an unrelated third
party suffer either the costs or the benefits of the externality and its
control. The distribution of costs and benefits across political units
is another important perspective. How are various political entities
affected by air pollution and its control? What impact might this play
on the adoption of final policy? All of these distribution questions are
examined in the following section using simulations from the
environmental-economic model.

The incidence of the cost of abatement of the air pollution from a
power plant is difficult to determine precisely. The problem is similar
to calculating the incidence of a sales tax on electricity. A precise
solution requires knowledge of how individuals may change their behavior
in response to the higher price of electricity. If the only change in
behavior is a higher price, then the incidence falls upon the consumers
of the electricity and indirectly upon the consumers of goods produced

from the electricity. If there is no change in behavior including holding prices constant, then the burden of the resource costs fall upon the producer of the electricity, the owners of the utility. If prices are allowed to vary and people are allowed to adjust to these higher prices, however, the incidence depends upon how people change their behavior. A precise solution to the incidence problem would require in depth knowledge of the own-price elasticities, the cross elasticities, and the income effects across all the groups one is interested in following.

For the purposes of this study, a number of assumptions are made in order to obtain rough estimates of the incidence of abatement costs. The incidence of abatement costs across income groups is assumed to be proportional to income. Adjusting for income, blacks and whites are assumed to share the costs of air pollution control equally on a per capita basis. The residents served by the utility are assumed to bear the full burden of abatement costs. That is, the indirect burdens through higher prices of consumer goods produced with the electricity or even higher costs of electricity sold to other utilities are assumed to be negligible. Given these assumptions, it is relatively easy to calculate what different groups will have to pay in terms of abatement costs alone.

Whereas the distribution of most goods (such as electricity) are determined by the market, the distribution of air pollution damages are determined by the laws of nature and where one chooses to reside. Given that air pollution effects are relatively small per person compared to the costs of moving, it is reasonable to assume that any relocations by the population because of one power plant will be minor. With information about where each group is located, it is possible to calculate how much each group will suffer from air pollution damage. If land prices do not change, these air pollution damages can be used to calculate the incidence of each abatement technique. If land prices do change, part of the benefits will also be shared by landowners, in which case, it is difficult to calculate the final incidence of benefits.

An uncontrolled coal burning power plant located in downtown New Haven would impose no abatement costs. The air pollution from such a plant, on the other hand, would have serious effects on all income classes. These effects are not shared equally. Families below the poverty line suffer the greatest health losses. Low income families average .017 days lost per person per year whereas the rest of the population averages .012 days lost per person per year. Non-health damages are more difficult to assess. Poor families have almost twice the exposure to pollutants which cause non-health damages than do more wealthy families, but richer families have more property to be damaged.

Though some abatement techniques will reduce damages proportionately across all income groups, other techniques benefit certain income groups more. Table VIII-1 presents the distributional impacts of selected abatement techniques across income groups. Low income families regain more days lost than higher income families from controlling air pollution. Dispersion techniques, in particular, favor low income families. This progressive impact of air pollution control is probably somewhat mitigated by the likelihood that upper income groups receive greater non-health benefits because they probably own more damageable property per capita than do low income groups.

Table VIII-1

The Distribution of Air Pollution Health Effects Across Income Groups

Abatement Method	Annual Per Capita Health Losses[a] (days lost x 10^{-3})		
	Below Poverty	Middle Income	Above $15,000
No Controls	17.5 (9.4)	12.7 (67.0)	12.4 (23.6)
Beneficiation 3	10.9 (9.6)	7.8 (67.1)	7.5 (23.3)
Flue Gas Desulfurization	7.1 (10.7)	4.6 (68.0)	4.0 (21.4)
Beneficiation 3 Plus Flue Gas Desulfurization	5.0 (10.9)	3.2 (68.1)	2.7 (21.0)
Electrostatic Precipitator (99%)	15.5 (9.2)	11.4 (66.8)	11.4 (23.9)
Remote Siting: 5 km East	9.1 (7.3)	8.4 (66.0)	9.4 (26.7)
25 km East	6.1 (7.3)	5.6 (65.4)	6.5 (27.3)
75 km East	4.0 (6.9)	3.9 (66.5)	4.3 (26.6)
Higher Stack (40 meters)	9.8 (9.3)	8.4 (67.7)	9.1 (23.0)
Beneficiation 3 Plus Higher Stack (40 meters)	5.8 (9.2)	4.9 (66.2)	5.4 (25.9)

[a] Figures in parenthesis are the percentage of total health damages endured by each income class.

The disparity between black and white exposures to air pollution is surprisingly greater than the disparity between the poor and the rich. Without air pollution control, black individuals suffer over twice the negative health effects that the average white individual endures. Blacks receive higher dosages of air pollutants primarily because they are concentrated in central cities (near the emission sites). Because blacks lose more active days as a result of air pollution, they also gain more health days as a result of air pollution control. Emission control systems invariably reduce black health losses by more than twice white health losses. Dispersion methods reduce black health losses by more than three times white health losses. Unless whites value health more than blacks, blacks receive more health benefits than whites from all types of air pollution control. Remote siting, especially, is clearly in the interests of the centrally located black population. Though all these control techniques favor blacks, they, nonetheless, also serve to reduce average white health losses.

The distribution of non-health damages across racial groups is difficult to determine without detailed data about the material and vegetation assets each holds. Blacks are exposed to higher pollution dosages which would infer higher non-health damages ceterus paribus. However, it is quite likely that individual whites own more property which would be subject to damage than blacks. The final distribution of non-health damages is, therefore, indeterminate.

A sentiment often expressed when faced with externality problems is to make the producer of the externality suffer the resulting damages. For example, the pollution from the power plant

Table VIII-2

The Distribution of Air Pollution Health Effects Between Blacks and Whites

Abatement Method	Annual Per Capita Health Losses[a] (day lost x 10^{-3})	
	White	Black
No Controls	11.7 (83.7)	29.2 (16.3)
Beneficiation 3	7.1 (83.2)	18.3 (16.8)
Flue Gas Desulfurization	4.0 (80.2)	12.8 (19.8)
Beneficiation 3 Plus Flue Gas Desulfurization	2.9 (79.6)	9.1 (20.4)
Electrostatic Precipitator (99%)	10.6 (84.3)	25.4 (15.7)
Remote Siting: 5 km East	8.4 (89.4)	12.7 (10.6)
25 km East	5.7 (90.2)	7.9 (9.8)
75 km East	4.0 (92.3)	4.2 (7.7)
Higher Stack (40 meters)	8.2 (87.9)	14.4 (12.1)
Beneficiation 3 Plus Higher Stack (40 meters)	4.8 (87.7)	8.7 (12.3)

[a] Figures in parenthesis are share of total health losses of each race.

could be blown over the users of the electricity. In order to explore this proposal, the distribution of benefits between the users and non-users of electricity are examined for each abatement technique. Users of electricity are arbitrarily defined in this example as the towns served by the utility which would operate the electrical generating station. This classification is somewhat arbitrary because the electricity could be shared by neighboring utilities and because some of the electricity is used in making products enjoyed by citizens across the country. Nonetheless, defining the consumers of the electricity as the residents of the towns served by the utility, one finds that these residents absorb only 50% of the total health losses caused by an uncontrolled coal burning boiler. However, the individual users of the electricity are subject to one hundred times the health effects of more distant non-users.

Non-health damages are concentrated within the utility's customer service area far more than health damages. Only 16% of the non-health damages from an uncontrolled plant located in New Haven are suffered by non-consumers of the electricity. Families within the utility service area have 500 times the non-health damages of the families outside of the service area.

All the pollution control techniques examined in Table VIII-3 indicate that the individuals served by the utility have more to gain from air pollution abatement than their more distant neighbors. Coal beneficiation, for example, reduces health losses by .3 days per person for electricity consumers as compared to .005 days per person for non-consumers. On a per capita basis, a town has substantial incentives to clean up its own pollution. The incentives for these consumers, though,

Table VIII-3

Distribution of Air Pollution Effects Between the Users of Electricity and the Non-Users

Type of Pollution Control	Population Served by the Utility		Population Not Served by the Utility	
	Non-Health Damages	Health Losses	Non-Health Damages	Health Losses
No Control	5153 (84.4)	695.8 (51.0)	9 (15.6)	6.4 (49.0)
Beneficiation 3	2161 (83)	449.7 (54.0)	4 (17)	3.7 (46.0)
Flue Gas Desulfurization	3537 (87)	362.5 (73.8)	5 (13)	1.2 (26.2)
Beneficiation 3 Plus Flue Gas Desulfurization	1383 (87)	263.9 (77.8)	2 (13)	.7 (22.2)
Electrostatic Precipitator (99%)	1467 (76)	579.7 (47.1)	4 (24)	6.3 (52.9)
Remote Siting: 5 km East	1208 (50)	216.1 (23.7)	12 (50)	6.7 (76.3)
25 km East	93 (9.9)	38.6 (6.3)	8 (90.1)	5.5 (93.7)
75 km East	20 (3.5)	12.7 (3.0)	5 (96.5)	3.9 (97.0)
Higher Stack (40 meters)	1662 (63.9)	250.3 (27.5)	9 (36.1)	6.3 (72.5)
Beneficiation 3 Plus Higher Stack (40 meters)	698 (61.6)	159.8 (29.7)	4 (38.4)	3.6 (70.3)

[a] Figures in parenthesis are share of health and non-health damages split between users and non-users of electricity. Health losses are in terms of 10^{-3} days lost/year. Non-health losses are in terms of mills/year.

are not as great as the incentives for society as a whole.
Furthermore, these consumers have a preference for certain abatement
techniques which are not shared by society as a whole. Siting plants
in more distant locations is especially attractive from the viewpoint
of the users of the electricity. Moving the power plant just 5
kilometers east reduces their damages more than the most expensive
hardware (an FGD system). However, from the viewpoint of non-users of
the electricity, siting plants away from New Haven is far less
attractive. In fact, the individuals immediately adjacent to a remote
plant are far worse off than if there were no abatement activities at
all. Though the average distant individual may benefit from remote
siting, specific individuals will always be worse off. Distant
individuals receive more health benefits from hardware controls in
general than from dispersion. Sulfur control systems are especially
attractive to residents of distant towns. Whether one is within the
utility service area or not will heavily influence one's preference for
hardware versus dispersion abatement methods.

Another important dimension to the pollution control question is
the distribution of costs and benefits across political boundaries.
Connecticut is not the only state affected by the air pollution from a
coal burning plant in New Haven. As can be seen in Table VIII-4, only
two-thirds of all health effects occur in Connecticut. New York suffers
about 18% of the health losses and Massachusetts and New Jersey suffer
another 7% and 8%, respectively. Surrounding states should have a very
direct interest in the pollution control activities of the New Haven
utility.

Table VIII-4

The Distribution of Air Pollution Health Effects Across States[a]

Type of Pollution Control	CT	NY	MA	NJ	RI	Other States
No Controls	91.5 (59.6)	7.6 (18.7)	6.6 (7.0)	6.3 (8.1)	3.6 (1.1)	2.8 (5.5)
Beneficiation 3	58.3 (62.2)	4.4 (17.6)	3.8 (6.5)	3.6 (7.6)	2.1 (1.0)	1.6 (5.1)
Flue Gas Desulfurization	43.7 (79.2)	1.5 (9.9)	1.2 (3.6)	1.1 (4.1)	.7 (.6)	.5 (2.6)
Beneficiation 3 Plus Flue Gas Desulfurization	31.5 (82.5)	.9 (8.4)	.7 (3.0)	.7 (3.5)	.4 (2.1)	.3 (.5)
Electrostatic Precipitator (99%)	77.9 (56.2)	7.5 (18.2)	6.5 (7.6)	6.1 (8.8)	3.5 (1.2)	2.7 (8.0)
Remote Siting: 5 km East	44.3 (43.1)	6.7 (24.3)	6.9 (10.9)	6.3 (12.2)	3.6 (1.6)	2.6 (7.9)
25 km East	16.0 (23.1)	6.7 (36.3)	6.9 (16.2)	3.8 (10.9)	4.5 (3.0)	2.4 (10.5)
75 km East	8.0 (17.0)	3.3 (26.2)	6.7 (22.9)	3.0 (12.6)	7.0 (6.9)	2.3 (14.4)
Higher Stack (40 m)	41.0 (40.1)	7.6 (27.8)	6.5 (10.4)	6.2 (12.1)	3.6 (1.6)	2.8 (8.0)
Beneficiation 3 Plus Higher Stack (40 m)	25.4 (42.0)	4.3 (26.9)	3.7 (10.0)	3.5 (11.7)	2.1 (1.6)	1.6 (7.8)

[a] Numbers in parenthesis are percentage of total health effects in each state.

An interesting result in Table VIII-4 is that Connecticut and the neighboring states have different self interests with respect to the choice of abatement techniques. On the average, Connecticut residents benefit more from moving a plant just 5 kilometers out of New Haven than they do from any of the more expensive hardware abatement techniques. On the other hand, all other states benefit far more from flue gas desulfurization than from dispersion efforts. In fact, some states may feel distant siting is harmful. Rhode Island, for instance, suffers heavier damages as a result of the relocation of plants towards its western border. New York and New Jersey, on other hand, are made substantially better off. Neighboring states are thus divided about the benefits of abatement by dispersion.

Is the small increase in health effects in Rhode Island and Massachusetts of .003 and .001 per person, respectively, large enough to warrant discouraging distant siting as an abatement technique? The comparison in Table VIII-5 is between the most efficient technique with dispersion methods and the best techniques without dispersion methods for several dollar values of a day lost. Banning dispersion is a cost to society as a whole regardless of the value of life (see Figure VIII-A). Which states gain or lose as a result of banning dispersion techniques depends on the particular value of life chosen. For example, if a day is worth 15 dollars, all states but Connecticut gain from banning dispersion methods. Connecticut always loses because it must pay the higher costs of abatement due to a ban on dispersion. If the value of a day rises to 20 dollars, different technologies become more efficient and no state gains from banning dispersion techniques.

Given most values of a day lost, however, the loss to society of a

Table VIII-5

Changes of Equity and Efficiency Caused by Banning Dispersion[a]

Value of a Day Lost	Value to State of Ban on Dispersion (10^6 \$/year)						Loss to Society
	CT	NY	NJ	RI	MA	Other States	
0	− 1.26	−.001	−.000	−.000	−.000	−.000	1.28
5	− 1.88	−.006	−.002	−.000	.002	−.001	1.84
10	− 2.39	.34	.148	.028	.198	.133	1.76
15	− 2.75	.45	.213	.059	.299	.192	1.54
20	− 1.70	−.012	−.005	−.001	−.004	−.003	1.72
40	− 2.33	−.29	−.39	.03	.04	−.09	3.03
50	− 3.20	−.36	−.49	.04	.05	−.11	4.07
75	−11.33	2.84	.76	.26	1.33	.84	5.30
100	−11.72	1.30	.70	.57	1.71	1.04	6.40
250	− 1.80	−.03	−.01	−.02	−.08	−.06	2.00
500	− 2.91	−.83	−.95	.08	.11	−.20	4.70

[a] Connecticut suffers from a ban on dispersion regardless of the value of life because it pays for the more expensive hardware abatement techniques. The benefits to the other states depend upon the value of life because different values encourage the use of different abatement techniques (see Table VII-4).

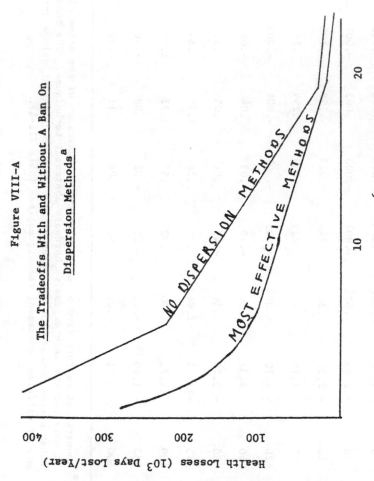

Figure VIII-A

The Tradeoffs With and Without A Ban On

Dispersion Methods[a]

NO DISPERSION METHODS

MOST EFFECTIVE METHODS

Health Losses (10^3 Days Lost/Year)

Non Health Costs (10^6 Dollars/Year)

a   Banning dispersion methods eliminates the cheapest available air pollution control techniques for every
    level of health effects.

ban on remote siting is larger than the gain of all the states who

benefit combined. That is, the states who lose from dispersion can be

compensated with the savings from the higher efficiency. For instance,

if the value of a day is 15 dollars, then society loses 1.54 million

dollars while all the benefitting states gain 1.2 million dollars. If

the value of a day is 40 dollars, then society loses 3.03 million

dollars while all the benefitting states gain only 40,000 dollars. Given

some values of life, redistributive effects of dispersion methods are

small compared to the large benefits to society as a whole. Without

a compensation scheme, states which are downwind of industrial centers

may find it in their interest to block remote siting as an air pollution

control technique even though it is efficient. Of course, states

which are upwind of the emissions or who must pay for emission control

have an incentive to site emission sources further downwind than is

efficient. The final resolution of this conflict of interests will

certainly depend upon the distribution of political power across states.

Hopefully, where efficiency plays a large role, the individual

differences among states will be ignored for the benefit of society at

large.

This analysis has generally assumed that the value of a day of

health is the same for individuals in every state. Because of variations

in age, taste, or income, all political units may not agree a life is

worth the same dollar amount. Of course, the analysis could easily be

redone assigning a different value of life to individuals in each state.

However, states often have an incentive to choose particular values of

life regardless of their real preferences. As can be seen in Table VIII-6,

every state but Connecticut would gain by stating their value of life is

Table VIII-6

Distribution of Air Pollution Costs Across States[a]

Value of a Day Lost	Type of Abatement	Total Dollar Loss From Air Pollution (10^6 $/year)[b]					
		CT	NY	NJ	RI	MA	Other States
0	Higher Stacks (40 m)	.40 (138.4)	.07 (95.8)	.03 (41.6)	.000 (5.6)	.03 (35.8)	.02 (27.9)
5	Higher Stacks (40 m)	1.13 (138.4)	.48 (95.8)	.20 (41.6)	.03 (5.6)	.17 (35.8)	.13 (27.9)
10	Remote Siting: 10 km	1.82 (97.1)	.77 (84.4)	.33 (37.0)	.05 (5.7)	.35 (37.8)	.22 (24.4)
15	15 km	2.38 (78.4)	1.15 (81.9)	.51 (37.0)	.09 (6.8)	.54 (38.7)	.33 (24.4)
20	Beneficiation 3 + Higher Stacks (40 m)	4.96 (85.7)	1.14 (54.9)	.49 (23.8)	.07 (3.2)	.42 (20.4)	.32 (15.9)
25	Beneficiation 3 + Higher Stacks (40 m)	5.46 (85.7)	1.44 (54.9)	.59 (23.8)	.08 (3.2)	.55 (20.4)	.40 (15.9)
50	Beneficiation 3 + Remote Siting: 25 km	6.60 (31.6)	2.45 (48.4)	.76 (14.5)	.21 (4.0)	1.10 (21.6)	.73 (13.9)
75	25 km	7.32 (31.6)	3.70 (48.4)	1.11 (14.5)	.30 (4.0)	1.65 (21.6)	1.06 (13.9)
100	75 km	8.51 (15.6)	2.41 (23.8)	1.15 (11.4)	.64 (6.3)	2.12 (20.9)	1.32 (13.1)
250	Beneficiation 3 + FGD + Higher Stacks	22.75 (15.2)	1.06 (4.2)	.44 (1.7)	.06 (.2)	.38 (1.5)	.27 (1.1)
500	Beneficiation 3 + FGD + Remote Siting: 25 km	23.71 (5.7)	3.07 (6.1)	.85 (1.7)	.26 (.5)	1.36 (2.7)	.79 (1.6)

FGD = Flue Gas Desulfurization

[a] This is the sum of abatement costs and air pollution damages from a single 500 MW power plant for New Haven. Connecticut is assumed to pay for all of the abatement. The damage burdens are calculated assuming that the most efficient abatement technology (given the value of health) is in place.

[b] The numbers in parenthesis are thousands of days lost given the technology in place.

250 dollars a day. This value minimizes the days of health loss from the New Haven power plant in each state. Assuming 250 dollars is far above the true value, the acceptance of this false value would provide a subsidy from Connecticut to all the other states and leave Connecticut to pay for all the loss of efficiency as well. Connecticut, on the other hand, has an incentive to understate its value of life because it pays for the abatement costs that everybody enjoys. At least with respect to this single plant, each state has an incentive to choose a particular value of life without regard to its true preferences. Generalizing this policy to sources from all the states will mitigate this bias. Nonetheless, states which produce relatively more or or less pollution may still have incentives to conceal their true preferences.

Footnotes

[1] See, for example, _Racial Discrimination in Economic Life_, Anthony
Pascal (ed.), Lexington, Mass.: Lexington Books, 1972; or
Race and Poverty, John Kain (ed.), Englewood Cliffs, N.J.: Prentice-
Hall, 1969.

CHAPTER IX

CONCLUSION

Pollution control technologies have frequently been assessed in terms of their abatement costs and their immediate impact on the environment. These initial impacts (tons of material released) are just links in a complex environmental system and though they may be indicative of relevant final effects, the immediate impacts have few implications for society in and of themselves. The rewards or benefits of air pollution control lie in the results at the end of the environmental chain, not in the beginning. Fewer illnesses, protected wildernesses, prolonged lifetimes, or increased agricultural production are the types of effects which are relevant to society. Past studies, however, have rarely assessed pollution control in terms of these final consequences. These studies fail to present or even grasp several critical tradeoffs in pollution control management.

This analysis improves upon past studies by assessing pollution control measures in terms of their relevant impact upon society: the resources utilized for abatement and the final consequences. In order to calculate these final effects, a wide range of scientific information is synthesized into an environmental-economic model. Through simulations of this model, it is possible to predict the expected result of various pollution control policies. The effectiveness, uncertainty, and distributional consequences of each policy can be estimated through these simulations.

This study also includes a detailed analysis of air pollution control options for a new 500 MW coal-fired electrical generating station for the city of New Haven, Connecticut. The numerical calculations are accurate only for a very narrow scope. Other sources may emit different pollutants and face alternative abatement choices and other locations may have unique meteorological conditions, spatial population distributions, and even air pollution abatement costs. The final numbers, then, are not readily transferable from the unique conditions studied here to more general situations. The process of arriving at these numbers, however, has broad applicability, not only to all sources of air pollution and all sites, but also to several other similar situations. This study's methodology can be used to study national air pollution control simply by repetitive studies across other sites and emission sources. The model is also applicable to studying water and soil pollution; the environmental scope of the model simply needs to be expanded. Finally, a number of health and safety issues can be better understood using the methodology demonstrated here. Wherever society is facing choices amongst preventative measures and environmental impacts, the type of analysis in this study can help illustrate the final tradeoffs implicit in the decisions.

The final tradeoffs in air pollution control are among abatement costs, uncertainty, health, material lifetimes, vegetation, and distributional effects. Though these tradeoffs are not understood precisely, there is, nonetheless, a considerable body of information available about the shape and nature of these relationships. This information has been incorporated into a computer based model in order to make this knowledge more accessible for policy purposes. Several abatement procedures are compared through simulations and a number of specific results are obtained. These results raise a number of important questions about current environmental policies.

The analysis of effectiveness in Chapter VII suggests that the Federal New Source Performance Standards lack effectiveness, at least in New Haven.[1] For instance, the standards require that a utility-sized furnace burning Northern Appalachian coal would have to remove 99% of its particulate emissions. Only high energy electrostatic precipitators, high energy scrubbers, and low air-to-cloth ratio fabric filters can remove such high percentages of particulates. The simulations indicate, however, that all particulate removal devices are ineffective in New Haven (see Figure VII-B). Additional expenditures upon high percentage removal techniques purchases very little. Table IX-1 demonstrates how high the marginal cost of a day saved is as a result of using these more expensive particulate removal devices. Installing such equipment implies that society values a day of health at over $500 or the life of a 35 year old at over $7,000,000.

Another hidden cost of current air pollution control comes from the strict regulation of new sources and the lax controls on existing sources. At least with power plants, old boilers require more fuel to generate a unit of electricity. They also emit more pollutants per unit of fuel burned. Placing the burden of pollution control on new units is counter productive because it encourages utilities to hold onto old boilers longer than they would otherwise. The improved thermal efficiency of new turbines alone can reduce emissions by between 10 and 50% if they replace older plants.[2] Also, new plants can be located at more desirable sites whereas the older plants tend to be concentrated near central cities. Discouraging new construction of power plants to replace older boilers can only serve to aggravate air pollution problems.

Table IX-1

Marginal Effectiveness of High Percentage Removal
Particulate Control Devices[a]

High Removal	Total Lost Days Avoided (10^3 days)	Total Cost (10^6 \$/year)	Marginal Cost of a Day Saved \$/day
Electrostatic Precipitator (99%)	49.9	13.68	2956.
Venturi Scrubber (High Energy)	53.9	13.78	734.
Air Bag Filter Air-to-Cloth Ratio: 2	63.3	11.01	556.

[a] Each of these devices is compared with other less expensive particulate control devices. The marginal effectiveness is the cost of the additional days gained.

Perhaps the most important result of the simulations in Chapter VII is that the price of ignoring dispersion techniques in New Haven is not insignificant. Both the New Source Performance Standards (NSPS) and federal ambient air standards provide no incentive for a polluter to utilize either higher stacks or remote siting. In order to conform with NSPS regulations, a new coal-fired power plant need not be built with any dispersion techniques, but it would need flue gas desulfurization and a high percentage removal particulate device. These techniques are expected to result in non-health damages and abatement costs of 26 million dollars a year and eliminate health losses of 378,000 days annually. If these same funds were applied to a combination of techniques which includes dispersion, over 44,000 more lost days could be saved. Similarly, if the 378,000 days saved is acceptable, another combination of techniques which include dispersion could save an expected $12,000,000 a year.[3] Existing legislation would cause millions of dollars a year to be wasted with this power plant alone primarily because dispersion techniques are not encouraged.

The NSPS regulations attempt to control pollution on the basis of the mass of certain emissions instead of the potential harm of the pollutants. The regulations neither recognize nor adapt to the substantial variations in local conditions across the country. The regulations cannot be effective across the wide spectrum of localities. Uniform emission standards are ineffective even within 100 kilometers of New Haven. As is evident in Table IX-2, the marginal damage from an emission of almost every pollutant varies dramatically across locations. A ton of each pollutant can cause between 5 and 30 times more damage if emitted in Rye, New York rather than New London, Connecticut. Any

Table IX-2

Marginal Damges of Emissions Across Three Sites[a]

($/Metric Ton Emitted)

Value of a Day ($/day)	Site: Rye, NY				
	SO_2	TSP	NO_2 [b]	CO	SO_4
			Site		
10	128.1	34.4	9.0	.4	612.4
25	278.7	55.4	7.9	1.0	1461.4
50	529.7	90.4	6.4	2.0	2876.4
100	1031.7	160.4	2.6	4.0	5706.4

	Site: New Haven				
	$SO2$	TSP	$NO2$ [b]	CO	$SO4$
10	69.0	9.5	9.0	.2	303.8
25	144.3	20.0	7.9	.5	728.5
50	269.8	37.5	6.1	1.0	1436.2
100	520.8	72.5	2.6	2.0	2851.7

	Site: New London, CT				
	$SO2$	TSP	NO_2 [b]	CO	$SO4$
10	29.7	1.2	9.0	.01	42.5
25	55.9	2.0	7.9	.03	85.6
50	98.7	3.2	6.1	.05	157.3
100	184.1	5.7	2.6	.10	300.8

[a] The value of visibility is assumed to be $.50 per kilometer person. Alternative values have only small impacts on total damages.

[b] Nitrogen dioxide damages are assumed to be qual across sites because the only known hazard is to vegetation and from acid rain which vary little across these sites.

uniform regulation which treats these emissions equally will either over control emissions from New London or under control emissions from Rye.

Uniform maximum ambient air regulations also ignore important variations in local conditions. The available scientific evidence suggests that dose response curves are linear in the range of typical dosages. The total health and material damages from a given ambient concentration of pollution will, therefore, be proportional to population density. A unit of a harmful pollutant in a dense central city area could easily cause 500 times more damage than an equivalent unit in a remote rural area. Uniform ambient air regulations will invariably either over control rural areas or under control central cities.

Though, in the long run, uniform ambient air standards are wasteful because they ignore local population densities, in the short run, they are probably more effective than uniform emission regulations. Most of the areas which violate the current ambient air standards are dense urban areas. Because the highest levels of pollution now occur in the most harmful places, lowering just the highest levels of pollution is reasonably effective. Uniform emission control regulations, in contrast, dissipate pollution control funds across the entire country. At least in the short run, uniform ambient air regulations focus pollution control dollars on appropriate areas.

Because of equity considerations amongst firms, "equal treatment", policy makers may be constrained to use uniform regulations across all firms. If a uniform treatment is required, the choice between price and quantity regulation of emissions can be important.[4] The scientific evidence suggests that the marginal damage curve for emissions at any given location is horizontal whereas the marginal cost curves rise with

increased abatement and vary sharply across polluters (see Figure IX-A).
With several different sources in any location, a uniform tax equal to
marginal damages is more effective than uniform cutbacks of the emissions
from each firm. A uniform tax across diverse locations, however, need
not be more effective than uniform emission cutbacks because a single tax
will not equal marginal damages across local areas. Across diverse
locations and individual sources of pollution, the expected difference
between uniform quantity and price regulation is not nearly as important
as the expected difference between uniform and local regulations.

A regulator's preference for price or quantity controls is also a
function of the uncertainty he faces. Given a regulator's uncertainty,
Weitzman [3] shows that the benefit of regulating with prices over
quantities (Δp) is approximately equal to:

$$\Delta p \overset{\circ}{=} \frac{\sigma^2}{2C^{11^2}} \ (B^{11} + C^{11})$$

Where σ^2 is the variance around the marginal cost curve, C^{11} is the
second derivative of the cost curve, and B^{11} is the second derivative of
the benefit curve. Using the results of the simulations in Chapter VII
and assuming that health is valued at $25 a day, it is possible to
estimate each of these parameters. The second derivative of the marginal
cost curve is about $8000/effective ton, the second derivative of the
marginal benefit curve is near zero, and the variance of the marginal
cost curve is about $100,000,000 in the neighborhood where expected
marginal costs equal expected marginal benefits. The advantage of using
price over quantity regulations because of uncertainty is about $120,000
for this example.

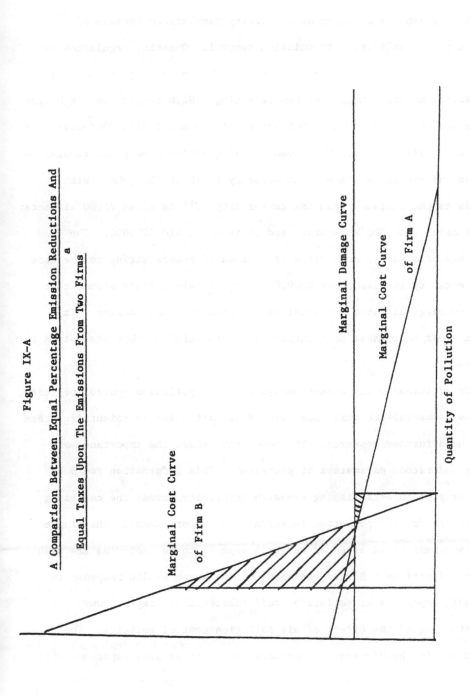

Figure IX-A

A Comparison Between Equal Percentage Emission Reductions And

Equal Taxes Upon The Emissions From Two Firms [a]

Marginal Cost Curve

of Firm B

Marginal Damage Curve

Marginal Cost Curve

of Firm A

Quantity of Pollution

[a] The inefficiency of uniform emission controls is equal to the hatched area. The hatched area under the marginal cost curve of Firm B is due to overcontrol of pollution and the hatched area above the marginal cost curve of Firm A is due to undercontrol of pollution with respect to the efficient level.

This advantage of price over quantity regulations because of uncertainty is only true for emission control. Quantity regulation is more effective with respect to uncertainty than price regulation when controlling emission height or remote siting. With regards to the height of the stacks, C^{11} is about \$1000/meter, B^{11} is about \$100,000/meter, and σ^2 is about \$10,000,000. The advantage of quantity over price regulation of stack height in the face of uncertainty is about \$500,000. With regards to the distance from the center city, C^{11} is about \$1000/kilometer B^{11} is about \$100,000/kilometer, and σ^2 is about \$10,000,000. The advantage of quantity over price regulation of remote siting in the face of uncertainty is also about \$500,000. These calculations assume that expected marginal costs are equal to expected marginal damages. It is not clear that these calculations are meaningful if this assumption is violated.

In addition to any direct analyses of air pollution control policies, the environmental-economic model also draws attention to scientific areas which need further research. The model can assess the importance of poorly understood parameters or processes. This information could be quite helpful in establishing research priorities across the competing disciplines for air pollution research funds. For example, the single most important broad areas of research appears to be improving the dose response functions. Better estimates of the human health response to low level exposures of pollution could substantially improve our understanding of the effect of air pollution control policies. The variations in the dispersion submodels appear to be less important with the possible exception of the rate of chemical transformation of sulfur dioxide to sulfate and long range transport. The range of engineering cost estimates fogs the distinction between relatively similar control techniques but the clarity is sufficient to distinguish several inferior methods. Improved engineering estimates, though helpful, are not as imperative as some of the other research items just mentioned. Though there is uncertainty about these estimates of uncertainty, certain areas of research stand out in need of further research funding.

As the scope of man's activities intensifies his impact upon the environment and as he further understands the nature of this impact, it is inevitable that he must face the implications of his actions. The need to comprehend and avail ourselves of the environmental options still open is of growing economic and political concern. We can ill afford to continue our ad hoc management of the environment. More effective, more equitable, and more reliable policies are available. Explicitly acknowledging the final implications of our environmental policies is an important step towards achieving a better environment in which to live.

Footnotes

[1] The New Source Performance Standards apply to all sizeable new sources which emit sulfur oxides, nitrogen oxides, hydrocarbons, carbon monoxide, or particulates. The standards limit the amount of each of these pollutants which can be released per unit of heat generated by a combustion source.

[2] See description of broad based abatement techniques in Chapter III.

[3] The more effective abatement methods obliquely referred to in the text include flue gas desulfurization, coal beneficiation 3, and either high stacks or remote siting.

[4] Price regulation refers to placing a tax or subsidy on a unit of emissions. The government can encourage desired behavior by adjusting the price of polluting. Quantity regulation refers to setting the level of emissions directly. The government can either tell a firm to reduce its emissions by so many tons or by a given percentage of its original emissions. The price and quantity controls illustrated here are a tax on emissions and a percentage reduction in emissions, respectively. For further discussion of price versus quantity controls, see Ruff [2] or Baumol and Oates [1].

Bibliography

1. Baumol W. and Oates, W. "The Use of Standards and Prices for Protection of the Environment." Swedish Journal of Economics, 73, (1971), 42-54.

2. Ruff, L. "The Economic Common Sense of Pollution." Economics of the Environment. Edited by R. Dorfman and N. Dorfman, New York: W.W. Norton & Co., Inc., 1972.

3. Weitzman, M. "Prices vs. Quantities." Review of Economic Studies, 41, (1974), 477-491.